Lecture Notes in Management and Industrial Engineering

Series editor

Adolfo López-Paredes, Valladolid, Spain

This bookseries provides a means for the dissemination of current theoretical and applied research in the areas of Industrial Engineering & Engineering Management. The latest methodological and computational advances that both researchers and practitioners can widely apply to solve new and classical problems in industries and organizations constitute a growing source of publications written for and by our readership.

The aim of this bookseries is to facilitate the dissemination of current research in the following topics:

- Strategy and Enterpreneurship
- Operations Research, Modelling and Simulation
- Logistics, Production and Information Systems
- Quality Management
- Product Management
- Sustainability and Ecoefficiency
- Industrial Marketing and Consumer Behavior
- Knowledge and Project Management
- Risk Management
- Service Systems
- Healthcare Management
- Human Factors and Ergonomics
- Emergencies and Disaster Management
- Education

More information about this series at http://www.springer.com/series/11786

Cesáreo Hernández
Editor

Advances in Management Engineering

Springer

Editor
Cesáreo Hernández
Management and Business Administration
University of Valladolid
Valladolid
Spain

ISSN 2198-0772 ISSN 2198-0780 (electronic)
Lecture Notes in Management and Industrial Engineering
ISBN 978-3-319-55888-2 ISBN 978-3-319-55889-9 (eBook)
DOI 10.1007/978-3-319-55889-9

Library of Congress Control Number: 2017935821

Printed on acid-free paper

This Springer imprint is published by Springer Nature
The registered company is Springer International Publishing AG
The registered company address is: Gewerbestrasse 11, 6330 Cham, Switzerland

Preface

This book collects the contributions presented at The New Industrial Organization Workshop, 2015, Valladolid, Spain. It gathered three generations of academics, many of them teaching Economics and Management in Schools of Industrial Engineering in Spanish Universities. It was an undeserved and unexpected tribute to celebrate my retirement as Professor Emeritus of the University of Valladolid. I ought to express my surprise, then, and my gratitude forever to all of them. The topics and the contents of the talks were free and they represent a collective view of Management with an Engineering focus.

Recently, the Industrial Engineering Standards in Europe (IESE) Project defines Industrial Engineering as "The branch of engineering that engages in the study of how to describe, evaluate, design, modify, control and improve the performance of complex systems, viewed over time and within their relative context". The definition shows the evolution of the field, in scope and focus, far beyond industrial activities, to cover any physical landscape populated by human agents and brings at the forefront the complexity inherent to social agents and their interactions. The focus has changed from product design and manufacturing until the eighties to developing distinctive capabilities of the firm and since the beginning of this century, collaborative management has been replacing management for competition.

These changes have occurred in parallel with changes in information technologies and computational capacity. Simulation in Economics (Artificial Economics) and Computational Organization allows generating bottom-up models, as in the true experimental sciences, complementing the analytical methods at use in Economics and the Management Sciences. The reader may remember the nightmare to calculate the analytical expression of the probability distribution of order statistics, such as the maximum height of a wave to design a pier, or the "fat tails" in economic data. Nowadays to obtain the probability distribution of order statistics, one can run Monte Carlo simulations (for example with Crystal Ball); as simple as this. Even more, these changes are leveraged by sharing knowledge in a free open access way such as the R Project for Statistical Computing or Creative Commons licenses free of charge to the public. Management Engineering can and should provide the skills and methods to manage complexity not limited to Industrial applications.

We have organized the chapters in four themes. I: Management Engineering and Organizational Sciences. II: The Governance System of the Firm. III: Heuristics, Intelligent Systems and Agent-Based Modelling. IV: Innovative teaching activities in Management Engineering.

In Chap. 1, the author presents a personal view of the evolution and classification of Management and Organizational Sciences. The contribution is very original and it is quite different from the usual academic papers. He has a wide and rich experience as a professor in Business and Economics, in managing public international and national institutions as well as in innovation and technology transfer at the Universitat Politécnica de Catalunya. He is Vice-President of the CYD Foundation and responsible of the yearly published report about Knowledge and Development in Spain. Although there are general principles in Management, he argues that the level and the type of institution determine a range of specific skills and competences. The reader will find that his ideas are well defined and have great practical and academic value.

No account is available, to our knowledge, of the origins and developments of Management Engineering in Spain and its relationship with similar studies in other countries. The authors are a good representation of the senior and young generations of experts in Industrial Organization (I.O). In Spain and in many Latino American countries, I.O is a mix of Technology and Management and it is the main tributary of today's Management Engineering. In Chap. 2, they review the historical background of Spanish Industrial Engineering and its relationship with the corresponding field in USA and in other countries. This chapter covers missing information about the actions undertaken by the first and successive Professors of the Chair Groups that stemmed from the initial "Economics, Organization and Legislation" of the Higher Technical Schools of Industrial Engineers. They end with the chapter presenting ADINGOR (Association for the Development of Organizational Engineering) founded to promote graduates' professional activity, international publications and congresses to gain visibility, helping to consolidate the Management Engineering field.

Economics and Industrial Organization are contextual and evolve. For this reason in Chap. 3, the author looks at the most relevant milestones of a long period of 60 years. The choice, the questions and the conclusions are personal and probably controversial. The chapter starts with the great failure of the Economy: the gap in the distribution of wealth even in developed countries. It continues asking the following questions: What have we learned from the seventies crisis? What are we learning from the current crisis? What was wrong with Economics as a social science? Can Experimental Economics allow us to understand and accommodate the social complexity of the Economy? What is the scope of Artificial Economics? Finally, since Artificial Economics provides solutions to complex problems, can we export socially inspired methods to other areas of Management Engineering? He concludes that there are tools to improve Economics and Management. However, to resolve the current economic and managerial challenges will require changes in methods and institutions that go beyond Economic Policy and the conventional Industrial Organization. The changes must be rapid and institutional. Not so much improvements in Economic Policy or Operational Methods as changes in Political Economy and Management Engineering.

Although the Economy and the Firm have common goals, Economic Theory has focused on markets and prices; dominated by a neoclassical view, it ignores the complexity of both, the Economy and the Firm. For this reason, Economics provided little help to managers. This fact prompted the raise of the new fields of Industrial Organization and Managerial Economics. In Chap. 4, the authors describe the dimensions of a New Industrial Organization that could help managing the Firm. They point out that the economic activity takes two alternative forms: the Market and the Firm. Economics is a social science that tries to explain how wealth is generated and how it is distributed. The Firm is a social organization whose members decide to cooperate to generate wealth and how it is distributed among the stakeholders. They certainly share a common goal. However, to translate economic principles to management is an open challenge. The traditional I.O, as it is understood among economists, deals with the generation of wealth through a market that, when it is well designed, achieves a fair distribution through endogenous dynamics towards equilibrium. On the other hand, a proper Theory of the Firm needs explicit rules of governance and operations. This fact requires a New I.O dealing with: Uncertainty far beyond probability; individual and collective bounded rational agents; specialization and heterogeneity; imperfect information and variety; incentives and penalties to avoid free riding; and how to develop core competences such as entrepreneurship, innovation and knowledge management. The chapter ends with a map of Management Sciences at the service of the New I.O in the sense of the classification presented in the first chapter.

Management Engineering is concerned with the design of governance systems for an agile and resilient firm. Chapter 5 deals with advances in organizational resilience and it provides a new systemic framework to analyze and design resilient organizations for firms involved in industrial or any other activity. Several works have proposed principles that one should follow to develop resilient organizations and the characteristics that a resilient organization should have. However, after reviewing more than 200 papers, the authors have found that these approaches lack a formal framework to create resilient organizations with capacity of creativity and innovation. A Viable System is a system organized in a way that it is able to survive despite changes in its environment. The theory of Viable Systems (TVS) is a scientific approach based on Organizational Cybernetics applied to the design and study of organizations and its processes. Could the theory of VS provide a formal framework to create resilient organizations? To answer this question the authors map the characteristics of resilient organizations to the characteristics of Viable Systems. They show how the TVS has conceptual and practical tools to design and monitor resilient organizations. They conclude that the TVS provides a formal framework to create resilient organizations.

Two core ideas in the Governance of the Firm are Agency Management in the Economic Theory and the Supply Chain Management (SCM) in Production.

An Agency relationship occurs between consecutive supply chain links. This is an example of convergence towards Management Engineering. Even for a single firm, the Bull-whipped effect showed that a perspective based on local optimization significantly damaged the supply chain performance. SCM needs a systemic approach: “thinking globally, acting locally”. Chapter 6 is dedicated to the Supply Chain Management for the twenty-first century. International trade has been multiplied by four since the year 2001. This means that the cooperation among the agents in the supply chain network, demands a systemic approach. How can SCM be successful in this new complex scenario? The authors discuss these issues and propose a three-legged framework for deploying the systemic (holistic) perspective in the SCM: Education, Orchestration, and Methodology. Education involves the development of a learning model based on analyzing the organizational issues as a whole (breaking down the ‘divide and conquer’ paradigm). Orchestration refers to the design and implementation of systemic solutions in the real world, which considers theoretical foundations, technological tools, and collaborative practices. Last but not least, a methodology to support the decision-making processes while adopting the collaborative scheme is essential to design a system with the desired properties and to align incentives between the different supply chain nodes.

The fourth technological revolution is already here and will involve deep changes in Management Engineering that will extend its scope beyond industry, civil engineering or architecture. In Chap. 7, the authors deal with the effect of these changes in Project Management Methodologies. The new technological revolution is propelled by the development of cyber-physical systems and technologies like Internet of Things, Big Data, Cloud Computing, 3D printing, etc. Therefore, we will see an avalanche of projects to implement new business models, products and services in their companies. They analyze the main characteristics of these projects and they wonder about the appropriate methodologies and managerial styles to lead them. They argue that these projects are complex in nature, according to the current literature on project complexity and thus, classical project management approaches might be unsuitable for managing them. They suggest some clues to seek for new managerial styles, mainly in the literature concerning innovation and new product development and within the “Agile” approach.

The objective of Chap. 8 is to study the implications of knowledge, cooperative relations and business innovation for “The New Industrial Organization” presented and discuss in Chap. 4. From Solow’s seminal article on growth theory, it is known that there is an unexplained growth, labeled as Solow’s “residual”. The sources of the unexplained growth are the intangibles: entrepreneurship, knowledge, increasing capabilities, innovation and networking. Collective intelligence emerges from the firm’s internal and external networking; it is a leverage of the intangible factors effect on growth because trading is a nonzero sum game. The authors present the implications of these intangibles to understand the meaningful connection between Economics, Business Organization and Strategic Management. To this end, they select two intangible factors: knowledge and cooperative relations between the agents of the firm. They are at the basis of business profitability, through their impact on innovation. Innovation exists to convert knowledge into products and services that will attract the customers. Competitiveness emerges here, when those goods and services are unique and generate value for the customers, who are willing not only to acquire them, but also to pay the set price. The challenge consists in offering, today as ever, products and services that are new, so that the firm will not be marginalized and abandoned by the customers, which are nowadays, agents of the firm's network. Knowledge, together with the innovative capabilities of the managers and employees, open up an enormous space for the firm, which leads to continuous progress full of opportunities and future: an emerging (endogenous) firm's growth.

In a global world, firms understand that cooperation is a key for sustainable innovation. How do the institutional endowment and innovation strategies determine cooperation? Chapter 9 provides contributions to answer this question. Knowledge accumulated inside the firm requires external feedback if the firm wants to reinvent itself. Institutional endowment influences those cooperative processes, acting as a barrier or as leverage. When building balanced cooperation strategies, the extent to which firms make their contributions counts in the innovation race by using "the wind behind" them, and it will determine how sustainable performance is achieved. Understanding the impact of institutional endowment on innovation strategies might reduce causal ambiguity and therefore help firms to contextualize their innovation strategies. The aim of the authors is to contribute with a type of cooperative innovation strategy dependent on three institutional factors: culture, networking and the legal system.

Heuristics and Computational Organization (Agent-Based Models) are new ways to deal with the complexity of the firm. Chapter 10 deals with ongoing research of the authors for improving tools for Nawaz, Enscore & Ham (NHM) Based Heuristics on Permutation and Blocking Flow Shop Scheduling Problems. Heuristics are approaches to operations management problems such as Flow Shop Scheduling that are NP-hard. Determining the best sequence of n jobs processed on m machines in the same order will give us $(n!)^m$ possible sequences. The most efficient heuristics for this problem is the NEH Heuristic. The authors analyze the behaviour of the insertion procedure of the NEH because in some cases the procedure does not improve the solution's quality. This fact occurs specially in the NEH- based heuristics, proposed in the literature of the blocking flow shop problem. Based on their analysis, they recommend evaluating the sequence before and after the insertion phase in order to retain the best of both procedures.

Agent-Based Modelling (ABM) is one of the most active modelling paradigms in many fields to handle the complexity arising in physical landscapes populated by social agents. This is the case with the Economy, or the Firm. ABM provides great advances and it is becoming a core method in Management Engineering. In agent-based modelling, each entity identified in the target system is explicitly and individually represented as an agent, and the different interactions among the agents and the environment are represented as well in the model. As in any experiment, it is always relevant to study the properties of the emerging results and when possible find out the relationship between the ABM and its parametric version. Chapter 11 provides guidelines for the appropriate design of experiments using machine-learning techniques in the analysis of ABM. This approach allows us to use a supervised learning algorithm to fit the results of the ABM simulations with the parametric representation.

Although agent-based modeling brings opportunities and advantages to Management and Organizational Design, it is a new approach in the management context and it has an important drawback: How to achieve a balance between simplicity and realism? One of the advantages of substituting human subjects by artificial agents, in order to model and manage organizations, is that the agents' behaviour can be controlled and simplified. Chapter 12 provides answer to a core question in organizational design. How alternative organizational configurations influence the emergence of corporate culture? Taking one of the many accepted definitions of corporate culture, the authors build a model to simulate different organizational designs in terms of agents, agent's endowments and scenarios and generate corporate culture. They find that formal organizational configurations are more likely to favour the emergence of corporate culture than informal interactions. The chapter is a nice example of the power of ABM to solve core social questions.

Intelligent Transport Systems (ITS) are fundamental to enterprise competitiveness, especially in terms of the efficiency improvement of Logistics and Freight Transport. Although these areas are commonly associated with the private sector, the public sector has a significant role regarding regulation, and the planning and maintenance of public transport infrastructures. Chapter 13 deals with two innovative areas in which public and private sectors collaborate: gateway facilitation technologies and some projects focused on an innovative management of freight transport. The use and development of ITS can generate significant benefits in terms of transport efficiency, customer satisfaction and environmental sustainability. The involvement of public institutions would be desirable to promote a single ITS standard that would allow the use of these systems.

In the last two chapters, the authors describe recent innovative teaching developments in Management Engineering carried out by them. In Chap. 14, the authors present the Lean School developed by them in conjunction with an industry partner (Renault) to improve the capabilities of the engineering students and employees of companies located in the Castilla-León region. It is a good example of university–industry partnership, where students learn lean tools in a simulated manufacturing environment, and industrial companies interact with universities to help creating world-class engineers and make a significant difference in engineering education. The Lean School Lab is used successfully by both for research purposes and learning-by-doing tool for students and industry employees.

Finally, Chap. 15 is a review of the innovative teaching activities carried out by the authors at the School of Industrial Engineering of the University of Valladolid in the field of Management Engineering. The objective is to promote the process of learning existing paradigms in the field of business organization and also that students acquire and develop the specific and generic skills they will need in their future professional activity. They present a portfolio of innovative projects of different learning methodologies and tools for supporting teaching activities. The chapter ends with a summary table of the improvement activities, which describes the elements for each action-project. They provide indicators of impact on professional skills and abilities, as well as of the benefits of each action and the contexts of use within the training process.

Acknowledgement is due to all the authors. One of the authors, Cristina Ruiz Martín, deserves special mention for her great help in managing the editing of the book.

Valladolid, Spain
2017

Cesáreo Hernández

Contents

Contributors

Ramón Companys Departament d'Organització d'Empreses, DOE—EPSEB—Universitat Politècnica de Catalunya, Barcelona, Spain

Isabel Fernández Department of Business Administration, University of Oviedo Polytechnic School of Engineering, Gijón, Spain

David de la Fuente Department of Business Administration, University of Oviedo Polytechnic School of Engineering, Gijón, Spain

Jesús Galindo-Melero Department of Business Organization and Marketing and Market Research, Department of Business Organization, and Commercialization and Market Research, School of Industrial Engineering, University of Valladolid, Valladolid, Spain

José Manuel Galán Grupo INSISOC, Área de Organización de Empresas, Departamento de Ingeniería Civil, Escuela Politécnica Superior, Universidad de Burgos, Burgos, Spain

Cesar Gamez-Alcalde Departamento de Economía de La Empresa, Facultad de Ciencias Sociales y Jurídicas, Carlos III University, Getafe Madrid, Spain

Nazario García Department of Business Administration, University of Oviedo Polytechnic School of Engineering, Gijón, Spain

Ángel Manuel Gento-Municio Department of Business Organization, and Commercialization and Market Research, School of Industrial Engineering, University of Valladolid, Valladolid, Spain

Ángel Manuel Gento Departamento de Organización de Empresas y CIM, Escuela de Ingenieros Industriales, Universidad de Valladolid, Valladolid, Spain

Cesáreo Hernández INSISOC, Universidad de Valladolid, Valladolid, Spain

Moisés Javato-Martín Department of Business Organization and Marketing and Market Research, Department of Business Organization, and Commercialization and Market Research, School of Industrial Engineering, University of Valladolid, Valladolid, Spain

Fco-Cruz Lario Research Centre on Production Management and Engineering, Universitat Politècnica de València, Valencia, Spain

Adolfo López-Paredes Grupo INSISOC, Departamento de Organización de Empresas y CIM, Escuela de Ingenierías Industriales, Universidad de Valladolid, Valladolid, Spain

Inés Magdaleno Departamento de Organización de Empresas y CIM, Universidad de Valladolid, Valladolid, Spain

Miguel Ángel Manzanedo del Campo Departmento de Ingeniería Civil, Escuela Politécnica Superior, University of Burgos, Burgos, Spain

Natalia Martin-Cruz Grupo INSISOC, Departamento de Organización de Empresas y CIM, Facultad de Ciencias Económicas y Empresariales, Universidad de Valladolid, Valladolid, Spain

Ricardo del Olmo Martínez Grupo INSISOC, Departamento de Ingeniería Civil, Escuela Politécnica Superior, Universidad de Burgos, Burgos, Spain

Angel Ortiz Research Centre on Production Management and Engineering, Universitat Politècnica de València, Valencia, Spain

Carmen De Pablos-Heredero Departamento Economía de la Empresa (ADO), Economía Aplicada II y Fundamentos del Análisis Económico, Facultad de Ciencias Jurídicas y Sociales, Rey Juan Carlos University, Vicalvaro Madrid, Spain

Javier Pajares INSISOC, Universidad de Valladolid, Valladolid, Spain; INSISOC, Social Systems Engineering Center, University of Valladolid, Valladolid, Spain

José Antonio Pascual-Ruano Department of Business Organization, and Commercialization and Market Research, School of Industrial Engineering, University of Valladolid, Valladolid, Spain

María Pereda Grupo INSISOC, Área de Organización de Empresas, Departamento de Ingeniería Civil, Escuela Politécnica Superior, Universidad de Burgos, Burgos, Spain

Raúl Pino Department of Business Administration, University of Oviedo Polytechnic School of Engineering, Gijón, Spain

Raúl Poler Research Centre on Production Management and Engineering, Universitat Politècnica de València, Valencia, Spain

Borja Ponte Logistics Systems Dynamic Group, Cardiff Business School, Cardiff University, Cardiff, Wales

Marta Posada Departamento de Organización de Empresas y CIM, Universidad de Valladolid, Valladolid, Spain

David Poza INSISOC, Social Systems Engineering Center, University of Valladolid, Valladolid, Spain

José Manuel Pérez Rios INSISOC, Universidad de Valladolid, Valladolid, Spain

Alfonso Redondo-Castán Department of Business Organization and Commercialization and Market Research, School of Industrial Engineering, University of Valladolid, Valladolid, Spain

Imma Ribas Departament d'Organització d'Empreses, DOE—ETSEIB—Universitat Politècnica de Catalunya, Barcelona, Spain

Cristina Ruiz-Martin INSISOC, Universidad de Valladolid, Valladolid, Spain; Department of Systems and Computer Engineering, Carleton University, Ottawa, ON, Canada

José Ignacio Santos Grupo INSISOC, Área de Organización de Empresas, Departamento de Ingeniería Civil, Escuela Politécnica Superior, Universidad de Burgos, Burgos, Spain

Pedro Sanz-Angulo Department of Business Organization and Marketing and Market Research, School of Industrial Engineering, Department of Business Organization, and Commercialization and Market Research, School of Industrial Engineering, University of Valladolid, Valladolid, Spain

F. Solé Parellada Departament d'Organització d'Empreses, DOE—EPSEB—Universitat Politècnica de Catalunya, Barcelona, Spain

Lourdes Sáiz-Bárcena Departmento de Ingeniería Civil, Escuela Politécnica Superior, University of Burgos, Burgos, Spain

Eduardo Vicens-Salort Research Centre on Production Management and Engineering, Universitat Politècnica de València, Valencia, Spain

Felix Villafañez INSISOC, Social Systems Engineering Center, University of Valladolid, Valladolid, Spain

Gabiel Wainer INSISOC, Universidad de Valladolid, Valladolid, Spain; Department of Systems and Computer Engineering, Carleton University, Ottawa, ON, Canada

Juan José de Benito-Martín Department of Business Organization and Marketing and Market Research, Department of Business Organization, and Commercialization and Market Research, School of Industrial Engineering, University of Valladolid, Valladolid, Spain

Part I
Management Engineering and Organizational Sciences

Chapter 1
The Evolution and Classification of Management and Organizational Sciences. A Personal Interpretation

F. Solé Parellada

Abstract Management sciences are comprised of a set of disciplines that can be applied to different fields, whose contents may vary due to the particular characteristics of the organizations under study. For example, differences occur when management sciences are applied to public management. In each field, some of the theoretical approaches used in the disciplines change, and new models and technical vocabulary have been introduce into the literature. However, beyond the addition of some very specific subjects, the set of main disciplines remains practically unalterable, as well as their position on the functional map. Nevertheless, the contents and relative importance of the management sciences may vary over time. The appearance of innovative waves in the management sciences can be observed clearly in changes in the contents of management manuals and journals on the subject. Often these changes are substantial and we should reflect on this.

Keywords Management sciences · Organizational design · Management evolution · Emerging strategies · Public management

1 Introduction

In almost all of the social sciences disciplines, there is a constant, high flow of papers with contributions of varying quality. Sometimes, a new, ground-breaking academic contribution is made in the management field. At times, a magic word serves as the basis for new reflections. At other times, a revealing event indicates that a change is about to occur in the nature of observed reality. After this, the literature takes off, and a wave of models and new vocabulary are generated that shake up the stock of consolidated knowledge. Therefore, in the management sciences, new contributions often signal the end of models that had been irrefutable,

F. Solé Parellada (✉)
Departament d'Organització d'Empreses, DOE – EPSEB – Universitat Politècnica de Catalunya, BarcelonaTech, Gregorio Marañón 44-50, 08028, Barcelona, Spain
e-mail: pacosoleparellada@gmail.com

C. Hernández (ed.), *Advances in Management Engineering*, Lecture Notes in Management and Industrial Engineering, DOI 10.1007/978-3-319-55889-9_1

but can no longer explain all or part of contemporary phenomena satisfactorily. Sometimes, or in fact frequently, the wave of changes results in new specialized journals and in groups of supporters who gradually bring new knowledge to the existing stock, reorganize it, consolidate it, and, with time, manuals and publications return to a consensus.

Examples are the emergence of Mintzberg's theory of organizational designs, considerations of strategy by M. Porter, or the subsequent formulation of the theory of resources and capacities. Henry Etzkowitz' triple helix concept is an interesting example of a ground-breaking innovation. Of course, Etzkowitz' contribution, based on the ingenious idea of the triple helix, corresponds to the area of development rather than that of management. However, the idea of the triple or quadruple helix has had a considerable impact on university management. It is difficult to explain universities' business models or strategies without referring to this concept, particularly in the area of the third mission. This is an example of induced, incorporated ground-breaking ideas. To sum up, flashes of knowledge, some of great significance, others simple but ingenious and sound, produce doctrine and are incorporated in management manuals.

The mechanics of growth and adaptation that is commonly found in the management sciences could be considered similar to that in the literature on natural sciences. However, the origin of waves of change in the two main branches of science is generally not the same. In fact, despite the abundance of publications on the state-of-the-art, the nature of progress in the management sciences is an area on which little work has been done. In this paper, we reflect on this subject, although we do not aim to cover it completely.

2 An Interpretation of How Management Sciences Advance

One possible interpretation of how management sciences advance is as follows. The world of organizations, which is the object of the study, could be considered a static world explored by researchers who strive to discover its essence. Consequently, we can assume that the systems in which organizations work are constant. Hence, the influence of the systems over the organizations would be easy to predict. In other words, we would be studying an ecosystem governed by laws that we need to discover, and with internal relations organized in a certain way that is easy to standardize. Clearly, agents have decision-making capacity and may make mistakes; and the system is obviously subject to decisions. However, what is not so clear is that the rules of the game do not vary, and that the organization of the system is constant in nature. Reality shows that the object of management sciences is constantly changing for various reasons. Therefore, we cannot fully accept the traditional role of the researcher, which is that of discovering the nature of a phenomenon as a sum of small discoveries used in different interpretations (models or theories) of a constant fact in management sciences.

During a specific period, changes in the economic organization of a territory are the result of (a) microeconomic changes; (b) changes in the economic model; and/or (c) changes in the organization of the sectors of economic activity under study, in other words, industrial organization or industrial economics. Even changes in technologies that aid management, such as ICT, sensors, big data and connectivity, have a profound effect on the essence of the object of management disciplines. Consequently, as a result of the initial conditions, researchers must check which changes should be made in the traditional theoretical approach, reconsider the models, and add new vocabulary. The set of disciplines are in constant evolution, and the relevance of the tools that they offer also varies over time.

A paradigmatic example of change in the object of study in the management sciences is that of the consideration of innovation as the basis of organizations' competitiveness. This consideration cannot be interpreted as the "discovery" of something that earlier researchers had not "discovered" and that they had therefore overlooked. In other words, we cannot state that open innovation, for example, was a practice that previous researchers would have underrated due to a lack of insight. The reason for their supposed oversight is not ignorance of the phenomenon or a lack of instruments to discover it, but the fact that "open innovation" was not something that was either possible or significant in the past, for exogenous reasons such as the emergence of industrial sector management as normal practice, accompanied by computer simulation tools. Therefore, the idea is not, in this case, to change the theory for a contingent approach, but to add theoretical aspects to a profound change in the nature of the phenomenon under study.

However, it is true that certain organizational innovations or changes in business models create new branches and sub-branches of management practices. One example is franchises, which would not have arisen easily without an increase in customers' purchasing power and technological advances that enable specific, replicable management systems to be implemented.

To sum up, management sciences evolve not only because of the capacity of researchers, but also because of changes in the object of study, and changes in the organization of the economic and social framework in which organizations work.

3 Management Engineering and Public Management

Of course, the above comments about the evolution of general management models and vocabulary could be extended to other areas of study, such as public management. However, changes would not occur at the same pace or intensity in these areas. An analysis of the reasons for this difference in pace and intensity could reveal surprising information about the nature of public management, and the causes of some of its inefficiencies.

An observation of differences between areas of application could help us to learn more about the nature of management and its limitations. It could help us to better understand the characteristics of these differences, the framework in which they develop, and the reasons for delays in adopting new management models and implementing tools that are widely used in other areas. We will discover the reasons for some areas of inflexibility, and the reasons for differences in the speed and extent of changes. Furthermore, to discuss and reflect on the change in management models in the public sphere, we must first consider the organizational nature of the institution in question. Public management in city councils is not the same as that in universities or state administration. There are also considerable differences in the pace of change, depending on which management discipline is considered. In any case, it is not our intention to discuss this topic in depth here, but to reflect not so much on what changes, but on what remains the same, in other words, the structure of the disciplines. To sum up, we will find out more about management sciences.

In my academic life as professor of organization I have been obliged and curious to reflect on the principles of management in general and of public management in particular. At the same time, in my professional practice, much of my work has focused on tasks that are directly related to the management of universities, university foundations and other public bodies. Consequently, I have had the opportunity to compare my personal approaches to management resulting from reflection and research. The responsibilities of a position often lead to reflection on adaptations of models and instruments that are designed to guide the management of a company so that they can be used in the management of a public body. This reflection is particularly important when, a few hours later, you have to explain these adaptations to students or write about them in a paper. However, such reflection is not compulsory, and sometimes I have reached the conclusion that it is vocational. Many managers have no problem with the contradiction between what they teach or publish and what they do. I do not know if this is a good thing or a bad thing: I do not judge it. However, this curious schizophrenia is not specific to those who work in management sciences.

My involvement as director of the UNESCO Chair of Higher Education Management (CUDU) at the UPC also gave me the opportunity to work on the theoretical principles and the models of evolution required in different management disciplines, applied to the needs of public management. In addition, it helped me to understand how we can lower the barriers and implement the right incentives, whilst maintaining the essence of public service. Some people, who are probably more gifted in management than I am or are not part of academia, do not need to reflect on the reasons behind their actions. However, academics do have this obligation to reflect.

4 A New Proposal to Define the Main Areas of Management Sciences

For an academic, this often-invaluable opportunity deserves a specific study to define the areas of management in the public and the private arenas, and the characteristics of each one. As mentioned above, it would be interesting to identify the main areas in which to apply management; the nature of the differences; the reason for the obstacles, which are often cultural, that affect incentives; and the circumstances in which the usual instruments are not appropriate. Finally, it would be of interest to consider how to establish a set of indicators and a range of objectives that are relevant and, above all, workable. The contradictions between proposals and the essence of the area of application, or between instruments and their suitability in one field and in others, is still a challenge that must be met and that would resolve many unnecessary ideological and practical conflicts.

In an initial approach, we should first outline and sort the categories of management sciences, that is, define the structure. However, to sort something, we need criteria. Personally, I have found it very useful to divide the sciences of organization into three main areas: (a) senior management, (b) production and product, and (c) management of a company's operation itself. These three areas overlap and I do not believe that this approach defines them rigorously. However, I hope that the explanation below is enlightening and helpful to both public and private managers.

4.1 Senior Management Disciplines

The first area covers disciplines related to senior management. We propose four groups of senior management disciplines, which establish the boundaries of this area.

The first group is that of strategy and business models, which could also be called models of value creation, an expression that is particularly relevant in reference to public administrations. We could describe this group as the part of management sciences that deals with the proposals and objectives of the company or the institution. Clearly, there are two main motors that drive the two disciplines mentioned above as part of this group: the satisfaction of demand, that is, the users; and the satisfaction of supply, that is, those who have founded the organization, who are the owners and/or those that run the institution. This statement seems banal as it is so obvious. In practice, a consideration of the two motors is fundamental, particularly in terms of models of value creation.

Once we know where we want to go, what objectives we have, and how we will go about returning the resources to the institution, we need to know which organization will take charge of these tasks. This is the second group. It is about knowing "who will do what" and what relationships of leadership, collaboration and information there will be between each person or group of people. We are talking about organizational designs. Organizational designs are one of the factors that influence the success of an organization and one of the key components of business models or models of value creation.

The third group in this area is that of information systems (IS): disciplines that help to show that objectives have been met at each level of responsibility of the organization. These disciplines and their instruments are essential to senior management. They include, for example, accountancy. IS help senior managers to make decisions that enable objectives to be brought into line with reality, and are also relevant in the rendering of accounts, with all its consequences. It is often stated that the differences between a public and private organization are revealed in this area of consequences: in the consequences and the ability to react to them. In practice, the differences are less significant than they appear.

Finally, the fourth group is comprised of disciplines related to “management” in the strictest sense, that is, guidelines on the conduct of managers. Among other subjects, this area includes: leadership, time management and management skills. These disciplines correspond to abilities and competences that are essential to the management of an organization, whether it is a company, a non-profit organization or a government body.

4.2 Operation Management Disciplines

The second area to consider is the production of the product or service, that is, the “area of operations”. Disciplines in this area describe and help to define the model of value creation that the organization in question considers adequate to obtain sufficient revenue and the right production costs, or simply to provide the public service for which it was created.

This second area answers the question: how will we provide the service or offer the product so that it is efficient, effective and satisfies users? The idea is to meet demand as appropriate and as planned. This area includes a set of disciplines that should help us to process the inputs to reach the desired outputs, as well as disciplines that will help us to come up with new products, new processes, and better ways to offer them. A range of disciplines will help us to achieve this objective. Some of them could also be placed in the third area of disciplines that can improve the internal management of the company.

To classify the specific disciplines in this section, we have already stated that they deal with “operations”. Nevertheless, we can also use phrases such as “innovation process” that are studied in the organization of production. Listing them will help us to understand and define their boundaries. The most common disciplines in this area include new products, creativity, design, R&D, production, logistics and distribution, innovation management, costs, purchases, quality and project management. The operational management systems that currently constitute the most advanced disciplines focus particularly on value creation and customer satisfaction. In fact, a high level of value creation in operations is what enables the organization to have sufficient revenue and cut unnecessary costs, and thus obtain the required result and the satisfaction of customers or users. In general, this area is not highly considered in public management, with exceptions that usually enable the incorporation of ICTs.

4.3 Functional Management Disciplines

The third area to consider covers all of the disciplines that will help managers to improve their work in relation to the effective operation of the company.

By this point in the article, it is clear that the boundaries between the three areas cannot easily be determined in the proposed classification, and the disciplines overlap in each area or between areas. It is not our aim to provide a classification with strictly defined boundaries; in fact this could only be achieved if we provided an intellectual guide concerned with understanding the organization and the place that it occupies. In this third area, this difficulty in establishing exact boundaries between areas is even more evident. However, perhaps due to their age and the abundance of literature, disciplines in this area tend to be more consolidated. Four of them should be considered in particular: accountancy, finance, human resources and marketing.

Accountancy is in fact an information system (IS), and was developed as such. However, time and the fact that, in many cases, accountancy is the only available coded instrument mean that it is often used for other purposes in the organization and by external agents. For example, accounts are essential to provide information for others, and accountancy is a tool used to indicate that taxes have been paid. Accountancy is an example of a discipline that is used in company management. Accountancy straddles the first and the third group: it is an information system that is available to senior management, but is also an instrument of daily management. Other instruments such as the "scorecard" have these two functions, depending on the level at which they are used.

Human resources are a set of disciplines. We need only consider staff selection or training to understand the range of disciplines covered by human resources. In relation to the second area of our classification, which we have called the innovation process or product and production, the discipline of staff management in production is frequently independent of human resources, which considers the relationship of employees with the company. In production, the staff is managed in terms of its function as participants in the process.

Finance have the same characteristics as human resources, and some of the same characteristics as accountancy. Managing the financing of a company's assets according to short-term needs is one thing; trying to find resources to make new projects possible is another. In general, this discipline develops in a steady way, as part of the internal operation of the company, "as long as there are no notable imbalances" and it is not a priority of senior management. However, the situation changes when assets and liabilities do not balance; when the available capital and liabilities are imbalanced in the short-term; or when major investments are planned. At these times, as in those of new business or organizational ventures, finance become the concern of the first area of senior management. This discipline has a different focus in public management and private management. However, the short- or long-term consequences for the institution or for the country are quite similar.

In our approach, the nature of the contribution of marketing can be divided into two parts: one related to the organization of sales and payments, among other factors; and the other related to production, which is interesting from an organizational perspective. New products, design, logistics and distribution are all elements that we have included in the second area. However, marketing professionals may not agree with this classification. This same difficulty can be found in the consideration of elements that are shared by human resources and the management of human resources in production.

In general, organized training in the management sciences does not pay the same amount of attention to the three areas and their elements. In particular, little attention is paid to the second area, as it is considered a technical aspect and/or one that has already been covered. Some disciplines, such as technological strategy or the (imagined) management of innovation, which present vocabulary and models such as open innovation, are included almost by obligation. However, the contribution and challenges of this essential aspect of organizations is not presented rigorously, even though it is the part that generates the product or the service that is requested or required.

5 Conclusions

As far as I know, our personal approach to the evolution of management sciences and the classification of its disciplines appears to be new and useful. Through a process of comparison, it enables us to question the differences that exist, and to understand why some actions are possible and others are not, and the reasons for the differences in the effectiveness of instruments.

Reference

The ideas of this paper are the product of my experience in teaching, research and working in private and public management. For this reason, I have avoided to provide references to authors that the reader may consider will support or question my proposal. Surely, there are authors and colleagues that have influenced my views, but it should be unfair to reduce these influences to an academic list.

Chapter 2
An Approach to the Industrial Organization Engineering Background in Spain

Ramón Companys, Fco-Cruz Lario, Eduardo Vicens-Salort, Raúl Poler and Angel Ortiz

Abstract In this article we review the historic background of Spanish Industrial Engineering and briefly compare it with its equivalents in the USA and other countries, indicating similarities and differences. We present the actions taken in Spain that have consolidated Organizational Engineering. First, we describe the early history in the older Schools of Industrial Engineering. We follow reporting the new Industrial Organization speciality in the Degree in Industrial Engineering and the Second Cycle of Industrial Organization Engineering that extends until the end of the last century. We present the actual academic organization to adapt to the European Higher Education Area (EHEA) along with the impact that its adaptation has had on the new Degrees, Master Degrees and Postgraduate Courses. Finally, a short section deals with the Organizational Engineering Development Association (ADINGOR), given its importance for the visibility and consolidation of Spanish Organizational Engineering in Spain and elsewhere.

Keywords Industrial organization engineering · Industrial engineering · University degrees · Spain

---

R. Companys (✉)
Departament d'Organització d'Empreses, DOE—EPSEB—Universitat Politècnica de Catalunya, BarcelonaTech, Gregorio Marañón 44-50, 3rd Floor, 08028 Barcelona, Spain
e-mail: ramon.companys@upc.edu

F.-C. Lario · E. Vicens-Salort · R. Poler · A. Ortiz
Research Centre on Production Management and Engineering, Universitat Politècnica de València, Camino de Vera, s/n, 46022 Valencia, Spain
e-mail: fclario@cigip.upv.es

E. Vicens-Salort
e-mail: evicens@cigip.upv.es

R. Poler
e-mail: rpoler@cigip.upv.es

A. Ortiz
e-mail: aortiz@cigip.upv.es

C. Hernández (ed.), *Advances in Management Engineering*, Lecture Notes in Management and Industrial Engineering, DOI 10.1007/978-3-319-55889-9_2

1 Industrial Engineering in Spain. Early History

It is worth briefly reviewing the history of the relation between Spanish Industrial Organization Engineering and US Industrial Engineering.

Spanish Royal Decree, of 4 September 1850, created industrial teachings in Spain, along with the Degree in Industrial Engineering. In the introduction to the above Royal Decree, the then Spanish Minister of Trade, Training and Public Works stated:

> …that it was essential to create schools, among other relevant institutions, where students of industrial careers could receive the necessary training to excel in art, or to become perfect chemists, skilled mechanics, etc., is found….
> …Art. 59. Students of the higher school in the first section will receive first-class mechanical engineers. Those in the second section will receive first-class chemical engineers. Those who study both degrees will receive industrial engineers.

The Spanish Royal Order, of 24 March 1851, created the Schools of Industrial Engineering in Barcelona, Seville and Vergara (the one in Barcelona known as the "Barcelona Industrial School"), and that of 26 August 1851, created the School in Madrid known as the "Royal Industrial Institute", which was the only school authorised to teach higher levels. Between September and October 1851, these schools began to operate in Barcelona, Madrid, Seville and Vergara. In 1855 and 1860, they opened in Gijón and Valencia, respectively. University Industrial Engineering teachings in Spain began 165 years ago.

A few years after these studies came into being, all the Industrial Schools in Spain closed, except for the one in Barcelona, due to an economic crisis. It was not until 32 years later that the Bilbao School opened as a continuation of the Vergara School. The Madrid School reopened 2 years later.

The Spanish Industrial Engineer was the result of combining Mechanical Engineer and Chemical Engineer. Electrical Engineering came later as a different activity to engineering. In 1864, Maxwell presented the Electromagnetic Theory; the first electric dynamo start working in Spain in 1874, and the first electric power station dates back to 1882 in New York. The creation of an Electrical Engineering option in the Spanish Industrial Engineering Studies was proposed in 1899, but was rejected.

Notwithstanding, a photograph exists which pictures a group of students during the 1899 graduation, with a footnote that indicates Electromechanical Engineers (the grand-uncle of the eldest author graduated this year).

In order to place the creation of the degree with other relevant events, it is worth indicating that the Ecôle Centrale de Paris (with which Industrial Engineering Schools share certain concomitances) was created in 1829. In 1848 the Barcelona-Mataró railway line was opened, which along with the Madrid-Aranjuez one, were the first railway lines to operate in mainland Spain.[1] In 1854 the Battle of

[1]Before in 1837, a railway line worked in Cuba (Havana—Güines), at he time an ultramar Province of Spain.

Balaclava took place during the Crimean War which witnessed one of the last cavalry charges, as featured in several films. Zurich Polytechnic was created in 1855, the same year that the Paris Universal Exhibition was held. In 1863 Meade defeated Lee in the battle of Gettysburg, which was the year when the London Underground opened.

Since its beginnings, not all Spanish Industrial Engineering training was technical. The Study Syllabus of the 1850 Royal Decree for higher levels reads, "Industrial Economics and Legislation: daily lesson", and this subject is also found in the 1855 Royal Decree. A utopian view, no less, idyllically considered training at three levels (20 May 1855, Francisco de Luxan, the then Spanish Minister of Public Works and Transport, during the presentation prior to the Royal Decree, stated that):

> elementary schools where an honourable craftsman and hardworking apprentice in workshops, for art's sake, also acquire sure means to put into practice procedures and results… professional schools…to offer expert operators and skilled builders to workshops and factories.
> preparation of industrial engineering…has to produce teachers, skilful builders of machines, and learned directors of large-scale workshops and the biggest establishments…

This aspect was reinforced by the Moyano Law of 1857, which entrusted industrial engineers professional management capacities.

The Civil Service Decree of 18 September 1935, not abolished until the present day, which with the creation of Industrial Engineering attributions, indicated in its Preamble:

> the Industrial Engineer career…has fully responded to the purpose for which it was created…training engineers so they acquire a solid scientific and technological basis that allows specialisation…to provide our industry with…competent managers.

Article 1 continues:

> this Decree…confers [to the industrial engineers] the total capacity to plan, execute and direct all types of installations and works in the technical and industrial economics branches.

Article 2 reasserts:

> Industrial engineers are particularly capable of acting, conducting and leading all types of studies, works and organisms in the economic-industrial, statistical, social and occupational domain

This history shows that, since a long time, Management has been one of the key pillars of the Industrial Engineer career.

What is an Industrial Engineer? If we consider that Engineering is a creative profession, one interested in developing and applying scientific and technological knowledge to cover human society's requirements, in a context with physical, economic, human, political, legal and cultural limitations, then a series of consequences arise:

- This profession is not merely technical as it includes economic, human, political, legal and cultural factors.
- It is based on a body of basic knowledge that must mastered to be successful in one's profession.
- It is concerned about centring on the creation of products, processes and services that are useful to society.
- To be successful, an engineer needs to have a series of skills and attitudes, which are related to Management and Econmics.

2 Industrial Engineering. The USA and Other Countries

Industrial Engineering appeared later in the USA than it did in Spain, but it is worthwhile remembering its origins given its major influence over the years. In 1886, H.R. Towne presented to the ASME (American Society of Mechanical Engineers) a communication entitled "The engineer as economist", which proposed that managing production plants was as important as managing their engineering. As a result of the works of Taylor (1856–1917) and Gilbreth spouses (1868–1924 and 1878–1972), a different engineering activity was undertaken at the Penn State University (USA) under the name of Industrial Engineering, with the first degrees taught in 1908, although the teaching of these studies went on after World War II. Its core subjects comprised the proposals of Frederick W. Taylor on Scientific Organization of Work. Nowadays more than 100 universities in the USA offer the Degree in Industrial Engineering.

Even though there are many points in common between Spanish and US Industrial Engineers, there are also considerable differences, and the fact that names coincide creates considerable confusion. When translating "Industrial Engineering" into Spanish, it was necessary to adopt the term "*Organización Industrial*". A literal translation into English ("Industrial Organization") has, more often than not, its ambiguities. Certainly the other possibility, "Scientific Organization of Work", must appear over-pompous or restrictive as it refers to "Work".

In the USA, Industrial Engineering ("*Ingeniería de Organización Industrial*" in Spain) has become well consolidated in both research and teaching terms. For over half a century, degrees of Industrial Engineer have proliferated in US Universities; they are increasingly important and more and more Industrial Engineering Centres and Departments appeared. This development has taken place in parallel with, but without interfering with, Business Administration Degrees and Departments.

Moreover, according to the "Engineering Education and Practice in the United States" report published in 1985 by the North American National Research Council, five main US Engineering branches exist: Electric/Electronic, Mechanical, Civil, Industrial (Organization) and Chemical.

Later on, Organizational Engineering was created under the name "Industrial Engineering", in other English-speaking countries like Canada and Australia, as well as “Ingeneria Industrial” in Latin America. This branch of Engineering in Europe comes in various forms, but with a convergent approach: “*Génie Industriel*” in France; “*Ingenieria Gestionale*” in Italy; “*Wirtschaftsingenieurwesen*” in Germany and Austria. It is worth pointing out that in France viewing an engineer, as a manager is traditional (perhaps the original Hispano version was inspired by it). An example of such is Henry Fayol (1841–1925), a mining engineer who attended the École de Saint-Etienne. He studied a technical and management career in France and wrote the result of his reflections and experience in French. He was a pioneer in defining the tasks done in a company, identified the management tasks, and ensured that management could be, and indeed should be, taught to engineers.

3 Spanish Industrial Engineering and Industrial Organization Engineering in the Last 65 Years

This section is an extension of the Keynote talks in Spanish of Companys (2001) and Fons Boronat (2001).

In Plan 48 (1948) there was a single Chair in the Organizational Engineering domain, formally called the Chair Group and also known as “Economics, Organization and Legal Issues”. This Chair embraced all the areas of “Theoretical and Applied Political Economy”, and the “Spanish Economic Structure in relation to the World”, “Industrial Administrative and Labour Law”, “Industrial Psycho Techniques”, “Occupational Health and Hygiene”, “Business Economics and Production”, and “Accountancy System Structuring of Industrial Companies”.

After two entrance examinations, those students who passed studied a 6-year career. As part of this career, one part of the 5th year and virtually all the 6th year were destined to the subjects we know today as Industrial Organization, and were common to all branches, of which there were four in Barcelona: Mechanics, Electrical, Chemical and Textile. Some subjects taught in year 6 were “Methods Improvement”, “Time Studying”, “Health and Safety”, “Statistical Quality Control”, along with a series of subjects which today are included in Operations Management.

Madrid had Professors Castañeda Xerta and Sierra Andrés, while Barcelona had Professors De Orbaneja and Aragón. Bilbao had Professor Beascoechea Ariceta, while Terrassa had Professor Valero Vicente. Their later actions were to stand out in the creation of the Faculties of Economic Sciences of their corresponding universities; Professor Castañeda Xerta was named the first Dean of the Faculty of Economic Sciences at the Madrid Complutense University when this Faculty was created.

The *Juan de la Cierva* Board, whose objectives were Industrial Organization and Standardization, owed its promotion to Professor Fermín de la Sierra Andrés, a Professor of Economics at the Madrid ETSII Centre. He founded the Work Rationalization Institute (in charge of establishing studies and applications of the rational principles of improving productivity in industrial and technical areas) and he was its first Secretary in 1946; and the National Industrial Productivity Committee (in which all State organizations related with Economics participated), of which he was Secretary General.

In 1955, through the joint order of the Spanish Ministries of National Education and Industry, the Madrid Industrial Organization School was created to train professional management staff in the Production Management and Business Management, undeniably the work of Professor de la Sierra, whose wish was a competitive Spanish industry of excellence at all levels.

In parallel, Professor De Orbaneja, a professor since 1943 at the School in Barcelona, created the "*Escuela de Organización Industrial*" (Industrial Organization School) in 1948 to teach Management studies in Catalonia, which acted as a platform to receive new knowledge in this area. This was the logical consequence of the Organization and Economics Seminar that created the Chair of "Economics, Organization and Legislation" at the Barcelona Higher Polytechnic School of Industrial Engineers in 1944. He was also the President of the National Committee of Industrial Productivity in Catalonia.

Professor Valero Vicente, Professor since 1948 of the Terrassa Higher Polytechnic School of Industrial Textile Engineers, in collaboration with RASA, and sponsored by the University of Navarre, the former Navarre General Studies, founded and was the first Director of the "*Instituto de Estudios Superiores de la Empresa*" (Higher Business Education Studies) (IESE). In 1964, he started his Master's Degree in Business Administration (MBA) in close contact with the Harvard Business School.

Meanwhile, Professor Beascoechea Ariceta, a Professor since 1960 at the Bilbao Higher Polytechnic School of Industrial Engineering, undertook a highly prolific activity in the R&D area with more than 15 books and over 70 highly original and important articles published about: Models of Management, Products and Technology, Analysis of Companies, The Costs and Applications of Human Behaviour, among others. They were all key areas of Companies.

With the 1957 Plan, certain optional courses became specialities, with a new one, Energy Techniques, and the Economics Chair Group was divided into two: Economics and Administration. Administration became in charge of three areas: Production, Administration and Organization. Moreover, the Industrial Engineers Schools changed from Special Centres to Higher Technology Centres, while the former three (Madrid, Barcelona and Bilbao) changed to four as that of Terrassa was incorporated into the Industrial Engineers Schools. Plan 57 was short-lived and it is not well known what became of it, but the reduction in the length of the studies were essentially to the detriment of Industrial Organization subjects.

Notwithstanding, the 1950s proved fruitful for Industrial Organization. In Barcelona, Professor De Orbaneja created the "*Instituto de Economía de la Empresa*" (Business Administration Institute). In 1958, he founded the "*Escuela de Administración de Empresas*" (Business Management School) (BMS), similar to the "*Escuela de Organización Industrial*" in Madrid; the BMS had plenty of teachers who were also from the Engineers school and taught the Industrial

Organization subjects of the subsequent Plan 64. One of the courses, which lasted 1 year, offered by the BMS in the 1960s was called "Mathematical Management Techniques", whose Dean was Professor Torrens-Ibern. He never taught many students, but the subject continued nonstop for almost 20 years. There were also production management, personnel, commercial, administrative courses, etc., and even a computer studies one at the end of the 1960s, for which Professor Companys was Dean.

Professor Torrens-Ibern had created the Special J.A. de Artigas Sanz Chair of Theoretical and Applied Statistics and the Applied Statistics and Operations Research Seminar. This Seminar and/or Special Chair ran courses and conferences, and took support in the BMS for "mailings" and in the School's Assembly Hall for meetings. At the end of the 1950s, Professor Kaufman gave an Operations Research course. In January 1964 Professors Torrens-Ibern and Companys organised the Hispano-French Symposium of Modern Management Methods, which covered two basic themes: Investment Programming and Activity Programming.

According to Decree 3608/63 in December 1963, and with OECD's sponsorship, the Seville Higher Polytechnic School of Engineers reopened. In July 1965 the first visit of experts to this organization in Seville took place to deal with the new syllabus to be taught at this school. This special "OCDE" syllabus was approved in July 1967.

Works to start building began in August 1965, and teaching activities began in Pavilion L-1 a year later in September 1966. The School was officially opened in April 1967, and its first Head was José Mª de Amores Jiménez.

Then Plan 64 arrived with a 5-year cut instead of a 7-year one. Once again the new Administration Chair Group was divided into two: "Organization of Production" and "Business Administration". Specialities still included Energy Techniques, introduced into Plan 57 and, assumedly, to compensate the cuts made, we find the Industrial Organization speciality. What was considered at one time a success for Industrial Organization in Spain, could also be considered a reduction in broad Spanish Industrial Engineering knowledge, and core Industrial Organization subjects disappeared. It was at this time when Spanish Industrial Engineering moved away from American Industrial Engineering, except those who studied the Industrial Organization speciality.

The end result was that three Chair Groups came into being in the area of interest: Group XI "Economics" (in charge of areas of the Economics: Theory and Structure, Business Management, Law and Psychosociology); Group XII "Administration" (Accounting, Costs and Budgets, Organization, Corporate Policy, Commercial Management and Information Integration); Group XIII "Organization of Production" (Organization of Production and Operations Research). To the above Higher Polytechnic Schools of Industrial Engineering, those of Valencia, Gijón, Valladolid, Vigo, Zaragoza, Málaga, Logroño, etc., were soon to join them.

As a result of Plan 64, the "*Instituto Politécnico de Valencia*" (Valencia Polytechnic Institute) came into being (to later become the Polytechnic University of Valencia), with the Higher Technical School of Industrial Engineers at its heart, but with an experimental plan arranged into semesters (actually 4-month periods due to holiday periods). Since the very beginning, a Department of Industrial Organization came into being, which comprised the four Chairs; their first professors were Professor Carot, Professor Dalmau, Professor De Miguel and Assistant Professor Lario.

In Valencia as of the 1980s and from the Business Organization Department, training activities were organised for professionals and businesspeople in the Industrial Promotion Institute (IPI) of the Chamber of Commerce, Industry and Shipping in Project Planning, Statistical Quality Control, etc. In the 1990s, training activities were provided at Ford España S.L. in Almussafes (Valencia), which later became an MBA for Ford executives, sat jointly by Ford España, Anglia University and Polytechnic University of Valencia.

Halfway through the 1980s, and encouraged by Professor Nieto, the then Director of ETSII Valencia, the first MBA of a public Spanish university was set up, known as the Master in Economics and Industrial Business Management (MEGEI), which continued until 1992. It came into being to support the EGEI Association with its corresponding Board made up of businesspeople and representatives of public and private institutions. In order to train businesspeople and professionals with no university qualifications, the Valencia Club of Economis and Business Management (CEDE) and the Business Organization Department of the UPV launched the Master in Economics and Business Management (MEGE), which was in activity for ten annual programmes.

These actions taken by the Polytechnic Universities of Catalonia and Valencia, and the brief notes made of the Madrid Polytechnic University make us realise just how important teaching activities and technology transfer were in the Organizational Engineering domain since Plan 1948, and how they helped consolidate it and conferred it visibility as one of the main areas of Engineering.

4 The Second Cycle of Industrial Organization Engineering Degree

In 1992, after successive changes and amendments to the Syllabi of various Industrial Engineers Schools, the Degree in Industrial Organization Engineering (IOE) appeared as a second Cycle, as set out in Spanish Royal Decree 1401/1992, of 20 November (BOE no. 306, of 22 December 1992).

Its main objective was to train professionals capable of performing analyses, evaluations, designs and management in its four main approaches: Operations Management, Organizational Management, Economic Research and Strategic Management. Moreover, degree holders received relevant technological training in the areas of Automatics, Building, Electricity, Electronics, Mechanics, Chemistry, Environmental Technologies and Manufacturing Process Technologies.

The Degree in IOE prepared graduates to manage and direct production and services companies, and various types of public and private institutions (public administrations, Universities, NGOs, consultancy firms, etc.) and in all its areas: project management, logistics, quality, costs, purchases, processes, production, products, commercial, finances, innovation and technology management, natural environment, human resources, etc. This dual technical and management training provided graduates with an overview of the company and its surroundings, so essential for improving the results of companies in any sector.

The Second Cycle Degree in IOE was taught in 27 Schools of 21 Spanish Universities. The difference between the Degree in IOE and other degrees was that it provided more knowledge about: Manufacturing Technologies, Production Organization and Management, Information Systems, Decision-Making Techniques, Modelling and Simulating Systems, Optimisation Techniques, Risk Assessments, etc.

We insist on differentiating the aspects of the Degree in IOE in organizational aspects and decision making in design, implementing and operating operational systems and services, such as integrating the above aspects into the technologies employed and the widespread use of models as a tool to analyse, make decisions and their practical guidance. All this results in the investigation of real business world problems, solved thanks to the solid conceptual bases acquired.

5 The Degree and the Master in Industrial Organization Engineering

A more extense analysis of the actual degree in Industrial Organization in Spain is the Spanish article by Mula et al. (2012). In here, we deal also with the Master in Industrial Organization.

In the ANECA (2006) white paper, it was proposed creating a Degree in Industrial Organization Engineering (IOE) that did not exist in Spain. This name, however, did feature as a speciality in Industrial Engineering, created in 1964, even though the Second Cycle Degree in IOE came into being.

This proposal summarized the various models then cited in a Report, these being:

- A 4-year Degree (240 ECTS) with an End-of-Career Project (ECP) in the 8th semester.
- The proposed Degree was to have a similar duration and structure to German Technical University Degrees, but with more Basic Sciences contents, like North American Industrial Engineers.
- This Degree was intended to prepare students to undertake their professions in a technical project and management domain, and to also offer a solid basis to those graduates who wished to do further studies to make an academic or scientific career.

On 14 March 2006, the External Experts Committee selected the degrees to be adapted to Europe, which included the Degree in IOE.

In 2008, ADINGOR (the Organizational Engineering Development Association) published the requirements to verify the “Degree in IOE”. This document was forwarded to all Spanish University Rectors with Industrial Engineering Schools. It defined a minimum of credits in various blocks: Basic Training, Industrial Technologies, Information and Communication Technologies, Quantitative Methods and Organization of Production, Business Administration and Economic Engineering, the Industrial and Technological Area and End-of-Degree Project.

Nowadays in Spanish Universities, Organizational Engineering studies appear in both Undergraduate and Master Degrees. Undergraduate Degrees centre on general Industrial Organization Engineers training from both the Industrial Technologies and Industrial Organization perspectives. The objective of Master studies is to specialise on any of the main Industrial Organization Engineering approaches.

The principal objective of the Degree in IOE is to train graduates so they can analyse, model, design, implement and improve complex systems that comprise people, materials, money, information, machinery, technology and energy in order to offer products and services as soon as possible and with the best possible productivity, quality, reliability and efficiency. Graduates also receive technological training in the areas of Automatics, Building, Electricity, Electronics, Mechanics, Chemistry, Environmental Technologies and Manufacturing.

The Degree in IOE is offered in 22 Spanish Universities. Practically all the Universities that taught the Second Cycle of Industrial Organization have adopted the Degree and some new universities have appeared.

The Degree is arranged according to the Industrial Branch Pattern, with a first academic year studying Basic Sciences, a second studying Industrial Technologies, the third concentrates on Organizational Engineering, and the fourth academic year which, in some Schools, includes several blocks of optional courses.

The Master in IOE is being implemented less into Spanish Universities which, in some cases have considered it not all that necessary given the implementation of the Master in Industrial Engineering. We should notice that the main aim of Master Degrees is to specialise in any of the main Engineering approaches.

6 The Organizational Engineering Development Association (ADINGOR)

A fundamental role in launching and making Organizational Engineering visible in Spain has been played by ADINGOR. A meeting was organised at the 3rd Organizational Engineering Congress held in September 1999, to establish the “Organizational Engineering Development Association” (ADINGOR), during which its statutes were approved. This took place at the Barcelona Higher Polytechnic School of Industrial Engineers, attended to, and represented by fellow members of this School and those of Bilbao, Burgos, Gijón, Leganés, Madrid, Seville, Terrassa, Valencia, Valladolid, Vigo, etc. The proposal of the ADINGOR statutes was prepared by a group of developers from these Schools, who worked on the definition and consolidation of Organizational Engineering for several years.

A few years before, the Second Cycle Degree in Industrial Organization Engineering was approved. Its acceptance and implementation obtained several replies according to diverse Higher Technical Schools of Industrial Engineering and their corresponding Universities, even though they all implemented it. The result was that some teachers from the Business Organization area at the Higher Technical Schools of Industrial Engineering held several meetings to deal with the Business Organization knowledge area and the Industrial Organization Engineering because implementing this Degree was a priority as it was a service for society and Spanish industry. Reaching a consensus on the presence and actions of teachers from the Business Organization area at Higher Technical Schools of Industrial Engineering was essential to outline the Organizational Engineering profile.

From the very beginning, it was necessary to characterise Industrial Organization Engineering appeared, as with other Engineering careers, especially those in the Industrial branch (Electrics, Mechanics, Chemical, Electronic and Automatic Engineering, among others.), but also to identify relations of the Business Organization knowledge area with Organizational Engineering. Very quickly a consensus was reached that two professional and academic versions existed in the Business Organization area. The first was linked to business in general and was closely related to Business Administration and Management. The other focused more on the technology/company interface, especially industrial business (which clearly moved towards services with many physical and technology resources). In short, Industrial Organization Engineering was conceived and identified with the technology/company interface and paid a great deal of attention to Operations, but placed much priority on resources and technology in general, and on their efficient and efficacious use in particular.

A main goal for ADINGOR was reinforcing and consolidating Organizational Engineering's own approach from Business Organization and Engineering. Much hard work was done on the content and scientific basis of this new field of knowledge, Organizational Engineering. The creation of a body of knowledge began, a conceptual body, but above all an applied and technological one that would allow a new Engineering field to be accepted, Organizational Engineering, by recognising its relationship with the development of Industrial Engineering.

ADINGOR's Board of Directors agreed about certain actions to be taken, which were ratified by the General Assembly, so that the Organizational Engineering Knowledge Area would become visible and be acknowledged as such. The academic order of subjects, teacher's promotion and research in Spain is made within the context of Areas of Knowledge in detriment of course of transversality. The first step to take was to be get the approval of "Organizational Engineering" as a new Knowledge Area. Several Universities, including the Polytechnic Universities of Madrid and Valencia, approved and the Area was well approved by the Universities Council for these Universities.

Another of ADINGOR's main objectives was met: deciding on a point of encounter, a forum, where opinions, as well as academic and research experiences, could be exchanged, without forgetting the relational capital among group members. In short, making an organization (ADINGOR) available that could represent the teachers and professionals who recognized themselves as belonging to "Organizational Engineering", and promote their yearly Congress of Organizational Engineering.

Another very important issue has been that of adapting the Undergraduate, Master and Postgraduate Degrees in Engineering to the European Higher Education Area. In September 2003 during the Board of Directors meeting in Valladolid to hold the CIO'03 International Conference, a document about the proposal of a Degree in Industrial Organization Engineering was approved. The General Assembly delegated its review, improvement and definite document to the Board of Directors. To explain the document, informative meetings were organised with the successive Presidents of the Assembly of Directors from Higher Technical Schools of Industrial Engineers, the Conference of the Directors of University Schools of Industrial Technical Engineers and Polytechnic Schools, the Rectors of several Universities and the University Council Technical Teaching Sub-Committee members. The majority offered their support and approval. This allowed a proposal of the IOE Syllabus structure to converge.

However, challenges were faced not only in relation to the Second Cycle, the new Organizational Engineering Knowledge Area and the Degree in Industrial Organization Engineering since CIO'06 held in Valencia. Following the agreements reached during the General Assembly, the successive Organising Committees continued to work on the internationalisation of both CIO and ADINGOR. The 2007 Congress in Madrid was also the 11th Organizational Engineering Congress and the 1st International Conference on Industrial and Management Engineering, while that 2008 one held in Burgos was the 12th CIO and the 2nd International Conference on Industrial and Management Engineering. In both cases, the official languages were Spanish and English.

By making full use of the Madrid Congress, the Board of Directors held a meeting with top representatives of the Brazilian Production Engineering Association, ABEPRO, which is equivalent to the Spanish Organizational Engineering. During the National Assembly, an agreement was ratified to tighten relations with ABEPRO, to collaborate in the call of joint congresses, in publications and journals and in research activities.

7 Conclusions

The present article has reviewed the relevant actions undertaken in Spanish Organizational Engineering activity, which has been similar to that of American Industrial Engineering in terms of the training, skills and competences of their professionals. The present work also indicates how Spanish Organizational

Engineering has become consolidated through the Degree in Industrial Engineering (the Organization Engineering speciality), the Second Cycle of Industrial Organization Engineering and the current Degree in Industrial Organization Engineering and Master Degrees covering the Organizational Engineering field.

The actions undertaken by the first and successive Professors of the Chair Groups that stemmed from the initial "Economics, Organization and Legislation" of the Higher Technical School of Industrial Engineers, their graduates' professional activity, international publications and congresses, the visibility of R&D&I in this field, have all helped consolidate the Spanish Organization Engineering field.

References

ADINGOR. (2008). Documento de Requisitos para la verificación del título de "Grado en Ingeniería de Organización Industrial". Boletín de la Asociación para el Desarrollo de la Ingeniería de Organización.

ANECA. (2006). *Libro Blanco*. Título de Grado en Ing: de Organización Industrial.

Companys, R. (2001). *La Organización Industrial en la ETSII*. Sevilla: Congreso de Ingeniería de Organización.

Fons Boronat, J. M. (2001). *La Ingeniería de Organización: Una visión desde la Administració de Empresas*. Sevilla: Congreso de Ingeniería de Organización.

Mula, J., Díaz Madroñero, M., & Poler, R. (2012). Configuración del Grado en Ingeniería de Organización Industrial en las universidades españolas. *Dirección y Organización, 47,* 5–20.

Chapter 3
Sixty Years of Economics: Some Lessons for the Future

Cesáreo Hernández

Abstract Economics is contextual and evolves. For this reason, in this chapter, I have selected some relevant milestones of a long period of 60 years. The choice, the questions and the conclusions are personal and probably controversial. The chapter begins with the great failure of the Economy: the gap in the distribution of wealth even in developed countries. It continues asking the following questions: What have we learned from the seventies crisis? What are we learning from the current crisis? What is wrong with Economics as a social science? Can Experimental Economics allow us to understand and accommodate the social complexity of the Economy? What is the scope of Artificial Economics? Finally, since Artificial Economics provides solutions to complex problems, can we export socially inspired methods to other areas of Management Engineering? I conclude that there are tools to improve Economics and to help us in designing proper institutional frameworks. However, solving the actual economic challenges will require changes in methods and institutions far beyond Economic Policy. The changes must be institutional and can not be delayed. Not so much improvements in Economic Policy as changes in Political Economy.

Keywords Economics · Management engineering · Agent based models · Artificial economics · Expectations · Political economy

1 Introduction

One of the advantages of being retired is that you have the time and opportunity to look back without anger and with a historical perspective. Not only to recollect one's personal life, but the professional one as well. In my case, the academic life. This entails a wide scope of freedom and of context to answer the question: What

C. Hernández (✉)
INSISOC, Universidad de Valladolid, C/Paseo del Cauce 59, 47011 Valladolid, Spain
e-mail: cesareo@eii.uva.es

C. Hernández (ed.), *Advances in Management Engineering*, Lecture Notes in Management and Industrial Engineering, DOI 10.1007/978-3-319-55889-9_3

seems to me the most relevant landmarks in the evolution of the Economy along the last 60 years? In this chapter, I give a personal and probably controversial answer.

The chapter begins with the great failure of the Economy: the gap in the distribution of wealth even in developed countries. It continues asking the questions: What have we learned from a crisis of the seventies? What are we learning from the current crisis? What was wrong with Economics as a social science? Can Experimental Economics (EE) allow us to understand and accommodate the social complexity of the Economy? What is the scope of Artificial Economics (AE)?

Finally, since AE provides solutions to complex problems, can we export socially inspired methods to other areas of Management Engineering? I conclude that there are tools to improve Economics and to help us in designing proper institutional frameworks. However, overcoming the actual economic challenges will require changes in methods and institutions far beyond Economic Policy. The changes must be institutional and can not be delayed. Not so much improvements in Economic Policy as changes in Political Economy.

2 Growth and Inequality

2.1 Growth and Inequality: The Evidence

Economic Theory is concerned with two main issues: how to generate wealth and how to distribute it. If we look with the lenses of Economic History, we can see that the Economy is doing well on wealth generation but badly in its distribution. In the last century, poor countries multiply by 3 their per capita income, from poverty; but the rich ones multiply their per capita income by 6, from richness. The income distribution gap has increased.

In the rich countries, since the early eighties the per capita income gap is widening to a point of becoming a major threat to democracy. As Sachs (2010) warned for the U.S.A. "*Amazingly, the richest 1% of American households now has a higher net worth than the bottom 90%. The annual income of the richest 12,000 households is greater than that of the poorest 24 million households... The level of political corruption in America is staggering. Everything now is about money to run electoral campaigns, which have become incredibly expensive. The mid-term elections cost an estimated $4.5 billion, with most of the contributions coming from big corporations and rich contributors. These powerful forces, many of which operate anonymously under US law, are working relentlessly to defend those at the top of the income distribution.... If this continues, a third party will emerge, committed to cleaning up American politics and restoring a measure of decency and fairness. This, too, will take time. The political system is deeply skewed against challenges to the two incumbent parties. Yet the time for change will come*".

Thomas Piketty's *Capital in the Twenty-First Century* best-seller and the views of Nobel Prize economists such as Paul Krugman and Joseph Stiglitz published in major newspapers and magazines are increasingly influential, boosting inequality to the top of the political agenda in mainly developed countries. Even the IMF is very concerned with inequality because it hinders growth. In the IMF's flagship publications, last June, three of its top economists raise the question: Is neoliberalism oversold?

Data of Figs. 1 and 2 are at odds with free markets and marginalism as the pillars of a fair distribution of wealth. It seems that growth comes at the cost of a soaring inequality. What is wrong?

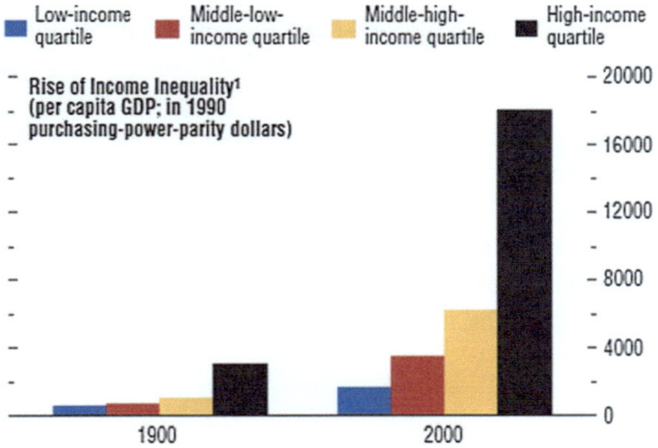

Fig. 1 Differentials in the world's income growth in the last century. *Source* IMF (World Economic Outlook, May 2000: Asset Prices and the Business Cycle)

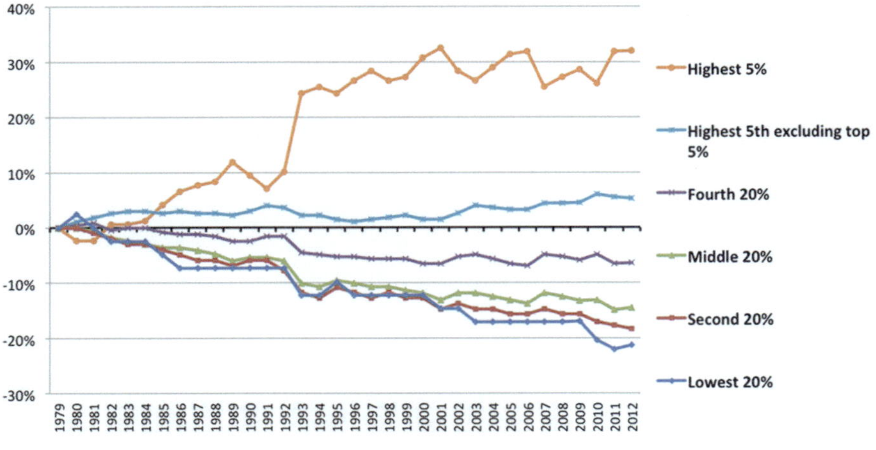

Fig. 2 Change in share of aggregate U.S Income since 1979. *Source* U.S. Census Bureau's Current Population Survey and MN2020

2.2 Trade as the Engine of Growth

Humans are the only living beings that can transform goods in an intentional way and can engage in trading. By transforming, specializing and trading, they generate wealth but not necessarily achieve its fair distribution. Figure 3 shows that goods export in 2015 is 100 times greater than in 1950. An average annual increasing rate of 6%. Yes, exchange correlates with growth.

Some of my students in engineering were quite surprised with the following example showing that the exchange can generate wealth even without physical inputs to the system: *The miracle of exchange*.

Imagine that there are two power plants that produce 8 Tm of SO_2. The regulator wants to reduce total emission to 4 Tm of SO_2 per day. There are two options: (a) each plant has to abate 2 Tm (b) they have to abate 4 Tm but the regulator leaves them to arrange how they clean 4 Tm. Table 1 shows the options and marginal cost of SO_2 abatement.

In the first option, the emissions abatement cost is 300 = 100 + 200 (Plant A); 900 = 300 + 600 (Plant B). Total cost with option 1: 300 + 900 = 1.200€. In the second option, plant A sells 1 permit to plant B for a price not lower than 300€ and abates 3 Tm. The plant B buys one permit to the plant A for a price not greater than 600 € and abates 1 Tm. Total abatement cost of option 2: 100 + 200 + 300 + 300 = 900€ with trade. Net value added by trading: 1200–900 = 300€.

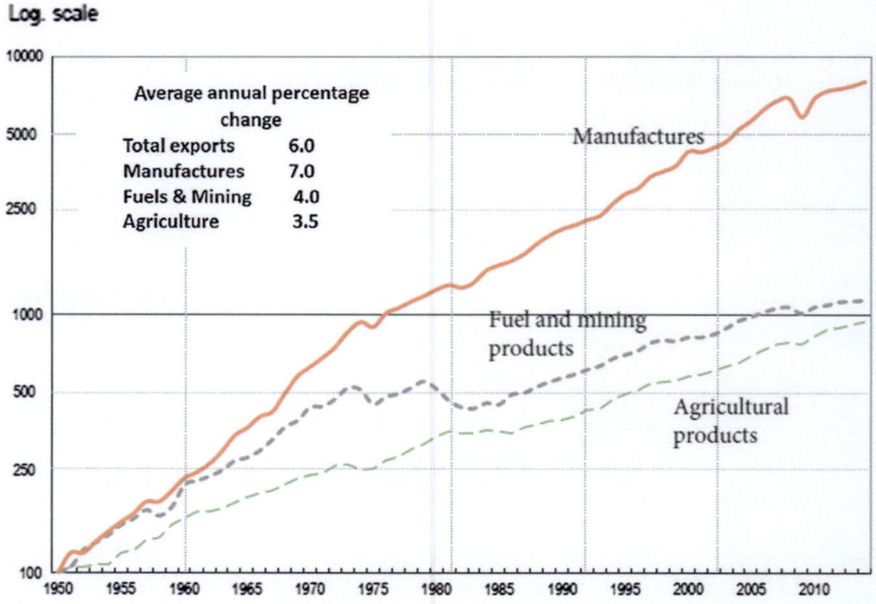

Fig. 3 World merchandise trade volume by product group. *Source* WTO

Table 1 Options and marginal cost of SO_2 abatement

Option 1			Option 2		
Abatement cost/Tm	Plant A (€)	Plant B (€)	Abatement cost/Tm	Plant A (€)	Plant B (€)
First	100	300	First	100	300
Second	200	600	Second	200	**600**
Third	**300**	**900**	Third	300	**900**
Fourth	**400**	**1200**	Fourth	**400**	**1200**

Any trading price between 300€ and 600€ will lead to an increase in social wealth of 300€. The market does not provide the fair transfer price. It is "mute" with respect to the actual permit price in this private personal exchange. Emission permits markets were created to allow for a collective strategy as the one described. They failed, mainly because the governments endorsed free permits to the participant firms, and/or the regulators intentionally design the market to induce clean production firms, as in the EPA Green House Gas (GHG) emission permits market Posada et al. (2007). The lesson to be learned from this simple example is that trading generates wealth but to achieve a distribution of this wealth one needs a properly design institution and a "visible hand".

A. Smith was well aware of this fact, and he advanced a proper answer: "*A regulation which enables those of the same trade to tax themselves, in order to provide for their poor, their sick, their widows and orphans, by giving them a common interest to manage, renders such assemblies necessary. An incorporation not only renders them necessary, but makes the act of the majority binding upon the whole. In a free trade, an effectual combination cannot be established but by the unanimous consent of every single trader, and it cannot last longer than every single trader continues of the same mind*" Smith (1776). The Wealth of Nations, Book I, Chapter X.

2.3 *The Solow's Residual: Celebrating the Instability and Change of the Economy*

In a seminal paper, Solow (1957) took a long time series sample, 1909–1949, of the USA economy to study the relationship between capital and labor with GDP. He found, an important unexplained residual. This empirical evidence came as a surprise. The rate of accumulation of income was not due to the sum of the accumulation of capital and labor, weighted by their relative shares in the production. There was a 40% of the growth rate unexplained. That is, a residual surplus after remunerate the factors according to their marginal contributions. Given the initial failure to identify the causes of this residual, it was attributed to technological change in a broad sense (techno capital, human capital, innovation, etc.). In an example of academic integrity, Abramovitz (1956) renamed it as "a measure of our ignorance".

Denote output by Y, labor by L and capital by K, the aggregate production function will be

$$Y = F(L; K) \times TFP$$

where TFP is the total factor productivity leverage produced by: Externalities, technological advances, learning, better knowledge and management improvements. Note the relevance of intangible factors such as I (institutions), M (management skills) and E (entrepreneurship). To express the residual in rates of change g, let W be the wage and P the price of the output. Then $g_Y = (WL/PY)g_L$ where WL/PY is the share (cost) of labor in output, denoted by a. Then we can write in rates of change

$$g_Y = a\ g_L \text{ and a similar way } g_Y = bg_K$$

In a general homogeneous production function, $Y = A\ L^aK^b$ the "residual" in rates will be

$$\text{residual} \ = g_Y - (ag_L + bg_K).$$

.

Once estimated a and b one can calculate the effect of increase rate of L and or K on Y and the residual.

Instead of using verbal accounting to explain the residual, an explosion of work was done on the role of several factors in determining growth and it was published in major economic journals trying to fit, not to explain, the residual. It was an outstanding example of bad scientific practice. If you wanted to progress in Mathematical Economics research at that time, it seems that following the flock of the academic establishment, was the easy way to publish in top rank economic journals. Of course, there were other economists undertaking research of true relevance with the focus in explaining endogenous growth as Romer (1990) and Lucas (1993). It is worth mentioning two core ideas that are derived from simple mathematics and creativity.

First idea. Flexibility in capital as a source of growth. Caballero and Lyons (1990) studied how increasing the number of intermediate factors for a given amount of capital will increase production.

Lets consider an homogeneous production function with constant scale economies with respect to labor L and capital K. What happens if K is divided into M intermediate capital factors x_i $i = 1 \ldots M$ with constant scale economies (CSE) as well?

$Q_0 = \ L^{1-a}K^a$ will be the initial production.

With the use of M intermediate factors, each one with the same amount of capital and with CSE we will have

$$Q_M = L^{1-a} \sum x_i^a = L^{1-a}M\ (K/M)^a = M^{1-a}\left(L^{1-a}K^a\right) = M^{1-a}Q_0$$

An increase in M, the variety of capital factors and maintaining the amount of L and K will increase production.

Second idea. An externality that is endogenous to the firm. Growth caused by knowledge (I + D) as a non-rival factor. Consider an example due to Romer (1990). I keep his numbers although learning in this industry has been exponential. A factory of hard disk drivers (hdd) for computers is using 1000 h of engineering to produce a hdd of 20 Mg. It actually there are 100 employees and a factory investment of $10 millions to produce 100,000 hdd/year = 2 trillion Mg/year.

Alternative one. The rival factors of the company are duplicated (factory and labor). Under the assumption of a constant return of scale production function, the output will double, to 4 trillion Mb of storage per year.

Alternative two. Suppose that the firm could have invested 20,000 h of engineering time (double engineering hours) in the design work instead of 10,000 h and, by doing so, could have a new hdd of 30-MB that could be manufactured with the same factory and workers. When the firm doubles all its inputs, it uses a 20,000-h design, 2 factories, and 200 workers and produces 6 trillion megabytes of storage per year, three times the original output.

Once the effect on growth of structural factors and intangibles was accepted, this "fitting" festival ended up with a narrative approach to establish competitiveness in terms of a mix of criteria. Many of them are intangibles factors, such as the quality of the institutions, the health care systems, the legal system etc. One of the most accepted narrative measures of growth capacity is the World Economic Forum competiveness' index. It is not by chance that the ranking in competitiveness is correlated with the index of transparency.

In view of these findings, how will be the shape of the firm's productivity curve? Since it constantly shifts up, the real productivity curve will never reach the decreasing zone. What does this instability imply? It implies that we are always in the decreasing zone of the mean cost function. Of course, in the ideal model, because the life cycle of a particular product will be too short. However, one can always think of a new product as a continuation of the old replaced product. Even for manufactured goods, this means that we approach a "Zero Marginal Cost Economy". Let us celebrate the instability and continuous change of the real Economy, which does not seem to obey the marginal rules! (Fig. 4).

The Earth is an "open system" receiving free solar "exogenous" energy. The Economy is a "social system", where collective intelligence and technological advances play the role of the sun. However, this source of growth is endogenous, maintaining endogenous development. Thermodynamic ideas are not proper for social systems. Solow's analysis means welcome instability of the Economy in the medium and long run.

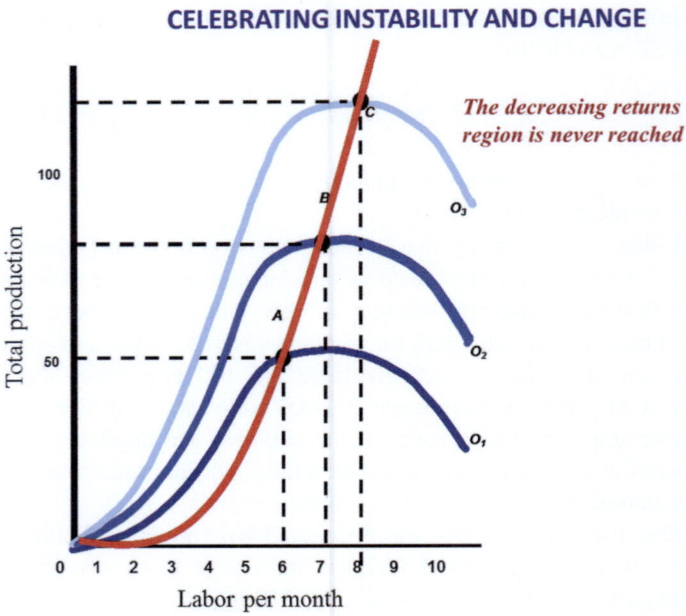

Fig. 4 Solow's residual, instability and change in Economics

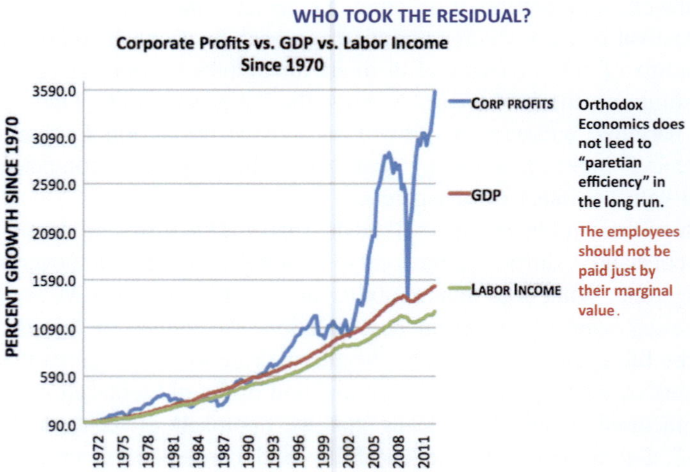

Fig. 5 The distribution of growth. Modified from Thompson (2013)

Solow's residual leverages growth; but does it brings equality? The answer is no. Figure 5 shows that corporate profits started to take off, relative to GDP growth, around 1985 and they soared before exploding in the last decade. It shows as well the short fall during the two first years of the financial crisis. What are the implications of this fact for growth itself and for inequality? We will answer this question later in the chapter. Now, we will concentrate in what happens with the non-financial activity.

2.4 *Preventing Inequality Needs to Reform the Social Contract (Rule of Law)*

Following the considerations above, there are three messages that economists and politicians ought to listen and citizens should be aware.

First message: The simplistic market-fundamentalist theories that have shaped Economic Policy during the last four decades are badly misleading, because GDP growth comes at the price of increasing inequality. Some symbols of reach citizens, as Buffet and G. Soros are already well aware of this message.

Second message: We need to rewrite the rules of the Economy and the social contract ("rule of law") to prevent the soaring rise of inequality and to ensure citizens benefit. It sounds like an indictment but failing to assume this challenge will be the politician's responsibility and will entail growth stagnation, a divided society and an undermined democracy.

Third message: Economist should never forget that progress with equality is not a question of Economic Policy measures but of Political Economy, such as the following ones.

(a) A reform of the social contract. The employees have to participate in the corporate profits, and of course, they have to share business risks as well. Wages based on productivity are not fair in the medium-long run, as the Solow's residual shows. Nothing new in the theories of firm's organization. For example, there are cooperative societies, sharecropping and similar contract arrangements in agriculture and in the fishing sector, since ages. Designing a new framework to accommodate particular types of business legal institutions is not a major problem. Institutional Economics is a well-established field in Management Sciences, where the firm is seen as a network of contracts. It is however a Political Economy question, beyond Economic Policy.

Unemployed people do not have wages nor can participate in the corporate profits, so we have to redesign the labor market to achieve full employment, with stable, yet when necessary, temporal jobs. The abuse of temporal unstable jobs is unacceptable. 63% of the part time jobs in Spain are people that do not find a full time job, according to Euro Stat. The Administration, certainly in Spain, is the main user of these bad practices. For example, Physicians with high qualification, with more than 11 years to graduate and achieving internship experience, can only get one-day contracts as duty doctors. The Court of Justice of the European Union has recently dictated sentence, declaring these practices at odds with the standing Directives, but so far nothing has been done to obey the sentence.

Ways to achieve full employment and decrease duality in the labor market are collaborative working, job-sharing and flexible work. Data of the level of job sharing is scarce but it is an increasing practice in the administration in UK, Switzerland and other advanced European countries. Another measure could be progressive relief contracts according to aging.

(b) Education. If a job is going to be shared, the employees have to be properly qualified. Lack of education is the main barrier to find a job. That means that education has to be recognized in the social contract (rule of law) as the main right of any citizen. It is an investment but at the same time favors equality. It has to be taken as a question of national interest to avoid the vicious circle: Poor people cannot afford education, therefore they cannot access to proper jobs, so they will be permanently poor (Fig. 6).

(c) Reforming the tax system. Education at all the levels and beyond the school and the university, needs financial support, or otherwise will be wishful thinking. This demands a serious reform of the tax system to increase the taxable income, now undermined by legal "tax avoision". The Economic Policy Institute (EPI) website has published facts to explain how the U.S. corporations rig the rules to dodge the taxes they owe over the last 20 years, Clemente et al. (2016). I quote three of the main facts of the report. The numbers may be different for other developed countries, but the need to increase the taxable income and the legal "tax avoision" practices is similar.

(i) Corporate profits are way up, and corporate taxes are way down. In 1952, corporate profits were 5.5% of the Economy, and corporate taxes were 5.9%. Today, corporate profits are 8.5% of the Economy, and corporate taxes are just 1.9% of GDP. (ii) Corporations used to contribute $1 out of every $3 in federal

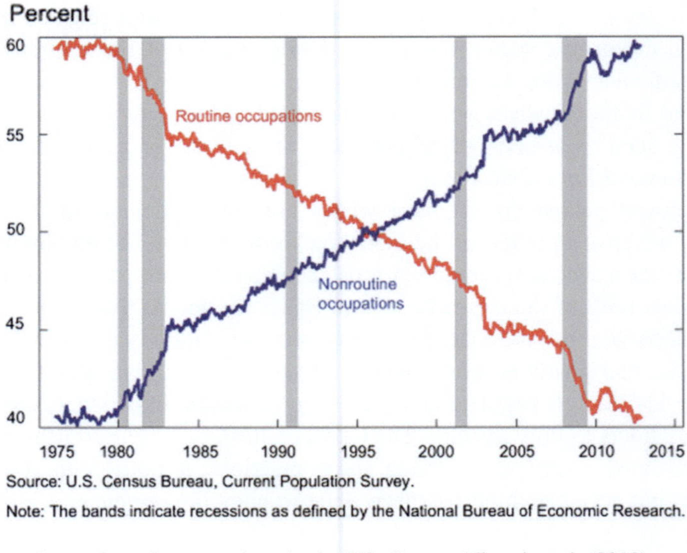

Fig. 6 Education and employment share in the US. *Source* Albensi et al. (2013)

revenue. Today, despite very high corporate profitability, it is \$1 out of every \$9. (iii) Many corporations pay an effective tax rate that is one-half (or less) of the official 35% tax rate.

The financial crisis brought attention not only to the problem of inequality, but also had a direct effect on it, as will be discussed later in the chapter.

3 What Have We Learned from the Seventies Crisis?

3.1 The "Augmented" Phillips Curve

As a student at the L.S.E, with a previous degree in electrical engineering, Phillips built an analogue machine using hydraulics to model the working of the UK economy, called MONIAC (Monetary National Income Analogue Computer). In the same vein, he thought that the excess demand and excess supply that restore the equilibrium in a commodity market could be applied to the wage rate and unemployment equilibrium. In Phillips (1958), he analysed the relationship between the rate of inflation and the rate of unemployment. With data for the U.K economy for each year, from 1861 to 1957, he found clear evidence of a negative relation between the rate on inflation and unemployment. "*Because of the strong curvature of the fitted relation in the region of low percentage unemployment, there will be a lower average rate of increase of wage rates if unemployment is held constant at a given level, than there will be if unemployment is allowed to fluctuate about that level. These conclusions are of course tentative. There is need for much more detailed research into the relations between unemployment, wage rates, prices and productivity*" (Fig. 7).

Fig. 7 Inflations and Unemployment in Australia during the 1970 crisis. *Source* Keen (2009)

The detailed research came soon. Samuelson and Solow replicated the Phillips's paper for the United States, using data from 1900 to 1960, and they confirmed his conclusions. They labelled the Phillips curve, and soon become the tool for Economic Policy decisions. It was simple and it was as robust and useful as a commodity market model. By implementing the right demand policies, governments hoped to achieve a permanent balance between employment and inflation assuring long-term GDP growth.

The 60s were a period of great moderation but with discrepancies between Keynesians and Monetarists over the Phillips curve and its consequences in terms of monetary and fiscal policies. Friedman (1968) and Phelps (1967) strongly disagreed. They argued that the apparent trade-off disappears if policy makers actually tried to exploit it—that is, if they tried to achieve low unemployment by accepting higher inflation. The agents cannot be fooled by the government unanticipated inflation eroding real wages (Fig. 8).

"Phillips wrote his article for a world in which everyone anticipated that nominal prices would be stable and in which that anticipation remained unshaken and immutable whatever happened to actual prices and wages....To state this conclusion differently, there is always a temporary trade-off between inflation and unemployment; there is no permanent trade-off. The temporary trade-off comes not from inflation per se, but from unanticipated inflation", Friedman (1968).

The Oil crisis changed the stable context; the use of monetary expansion generated more inflation with unemployment. Friedman made his contribution in 1968, before the Oil crisis and it is a good example of the Chicago's wisdom from discontent. However, he was not recognized by the academic establishment until the new context of the Oil crisis "experiment". Before the crisis, there were scenarios different to the UK and the USA data, but it was better for Keynesians and policy makers to keep the new toy, and to ignore these alternative contexts (Fig. 9).

Nevertheless, by the middle of the seventies, the expectations augmented Phillips curve was accepted and Monetarism wisdom as well. Rules versus

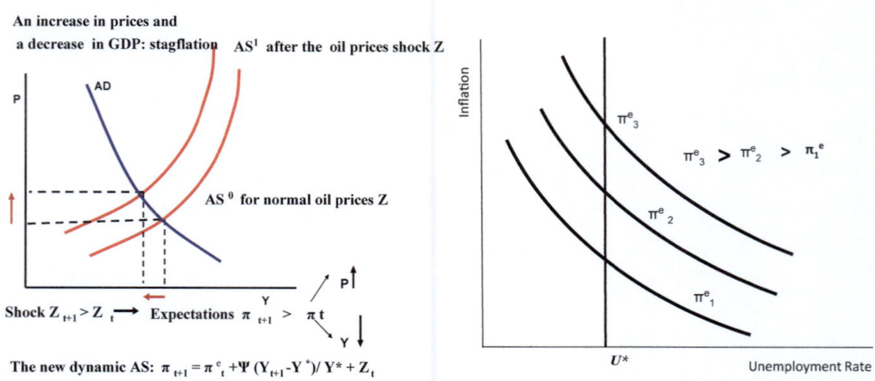

Fig. 8 The effect of oil prices and the "expectations augmented" Phillips Curve

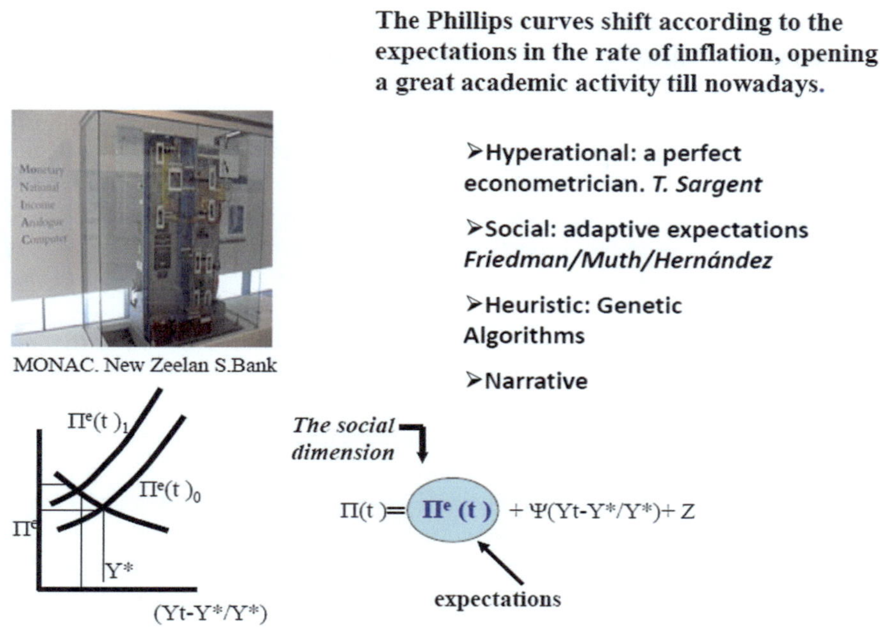

Fig. 9 The Phillips curve was so fashionable that even analog machines were built

discretion in economic monetary policy, was a consequence. The rate of change m should be determined from the target rate of inflation π and the expected rate of growth y: $m = \pi + \lambda\ y$.

3.2 *Rational Expectations*

The main contribution of Friedman and Phelps is clearly emphasized in the above Friedman's quotation. The temporary trade-off comes not from inflation per se, but from "unanticipated inflation" opening a new debate that still goes on. That the economic structural equations should include expectations was pointed by Keynes, and they were already incorporated in the consumption and investment functions, with narrative explanations: The life cycle by Modigliani, the permanent income by Friedman and extrapolative (memory) by Klein. However, in this crisis it affected the aggregated supply (AS).

How to model these expectations is an open question, and it is a reason why macroeconomic predictions of available models differ substantially in the medium and long run, Blanchard and Johnson (2013).

The recent crisis has demonstrated the inadequacy of models based on the assumption of rational expectations (RE) even with the last twist of what do we mean by RE. The theoretical guardians of RE and the thousands of users of econometric models remain silent in their trenches. The issues raised 40 years ago are still relevant today. Let us recall some definitions of RE proposed by distinguished protagonists of the debate.

Muth (1961) introduced the notion of RE: "*In order to explain these phenomena, I should like to suggest that expectations, since they are informed predictions of future events, are essentially the same as the predictions of the relevant economic theory*".

Sargent (1993) gives us an interpretation of this definition. "*The idea of rational expectations is ... said to embody the idea that economists and the agents they are modeling should be placed on an equal footing: The agents in the model should be able to forecast and profit-maximize and utility-maximize as well as the economist —or should we say econometrician—who constructed the model*".

He goes over himself to claim "*Rational expectations impose two requirements on economic models: Individual rationality, and mutual consistency of perceptions. When implemented numerically or econometrically, rational expectations models impute much more knowledge to the agents within the model (who use the equilibrium probability distributions in evaluating their Euler equations) than is possessed by an econometrician, who faces estimation and inference problems that the agents in the model have somehow solved. I interpret a proposal to build models with bounded 'rational agents' as a call to retreat from the second piece of rational expectations (mutual consistence of expectations) by expelling rational agents from our model environments and replacing them with 'artificially intelligent' agents who behave like econometricians. These 'econometricians' theorize, estimate and adapt in attempting to learn about probability distributions which, under rational expectations, they already know*".

It is difficult to grasp the message of this rhetorical masterpiece. It seems that following his numerous works, after using VAR time series models as a proxi for an econometrician's rationality (that is models without theory) he was interested in how the econometrician agent learned with artificial intelligent agents. In particular, if using recursive least squares (well known in identification systems in Control Theory) will converge to a (RE) equilibrium (Marcet and Sargent 1989).

The reader probably will share the following indictment: "*I am not well qualified to criticize the theory of rational expectations and the efficient market hypothesis because as a market participant I considered them so unrealistic that I never bothered to study them. That is an indictment in itself but I shall leave a detailed critique of these theories to others*" Soros (2012). One of the earliest critics was my paper Hernández-Iglesias and Hernández-Iglesias (1981). In what follows we present the fundamental arguments referring the interested reader to the article for the analytical details.

3.3 Rational and Adaptive Expectations

Three questions need a proper answer. How do agents form expectations? Could adaptive expectations (AEx) be equivalent to full rational expectations? Can fast and frugal individual predictions lead to aggregate social expectations that will also end up been rational?

I am glad to find out that some work I did in my early academic carrier may be useful 40 years later, to answer these questions. In the early seventies, I was a student in the PhD programme in the Control and Computing Department of the Imperial College (IC), London. The IC was involved in a quite new project to apply Control Theory to Econometrics (PREM) with the Queen Mary College as partner then, and the LBS later on, in what ended up as the Forecasting Unit. At the same time, my brother Feliciano just returned to Spain as a Ph.D. in Economics at the U. of Chicago. He was an adviser at the Spanish Ministry of Industry, which was the responsible of the Spanish Business Intentions Survey (SBIS). The National Statistics Institute mainly in hands of economists, wanted to take over the SBIS. I was in charge of a one-year project to assess the predictive capacity and the internal and external consistency of the SBIS with the time series methods of Box-Jenkins. The approach at the I.C was quite advanced and different to the B-J methods in several relevant issues. I set up a team with other two Ph.D. students and start developing our own suite of programs in Systems Identification (SYSID) with available knowledge in prediction errors methods and adaptative control that later on led to expectations, optimization methods (Ghahramani 1998) and prediction error methods (Ljung 1999). A short description of the results was published in Hernández et al. (1979).

Looking at the conundrum of rational expectations and at the shocking results in causality tests with economic time series, from this alternative engineering approach, could be a good novel idea. After debating with my brother for more than a year, I decided to stop deliberating and to send the paper for acceptance review to the editor of The Journal of Econometrics, Zellner. We had a very encouraging and friendly response with several deep remarks and some references to read, that were of great help, and the revised paper was published (Hernández-Iglesias and Hernández-Iglesias 1981). The paper has been compulsory reading in the post-graduate courses delivered by Zellner at the U. of Chicago for some years. It received attention in Europe, but given its disruptive nature and the disagreement with some of the visible leaders of the RE “revolution” in the US, we have no response to our indictments from them, except from Zellner and Cagan.

3.4 Expectations and Efficiency

At that time, expectations models in practice were linear distributed lagged models such as in Cagan (1956) studying the monetary dynamics of hyperinflation:

$\hat{y}_t(k) = \sum_{i=0}^{i=n} l_i(k) y_{t-i}$ where $\hat{y}_t(k)$ is the anticipated value from t for k periods ahead. Because multicolinearity is present, it was not possible to estimate the l_i weights and the extra condition of parsimony was imposed by assuming a geometric decay in the weights. As a "behavioral" alternative the expectations could be approximated by an error learning process $\hat{y}_t - \hat{y}_{t-1} = \alpha(\hat{y}_t - \hat{y}_{t-1}) = \alpha\hat{\varepsilon}_t$ of the prediction error $\hat{\varepsilon}t$. This approximation was efficient only if the univariate time series follows an ARMA (1, 1): $y_t - y_{t-1} = \varepsilon_t - (1 - \alpha)\varepsilon_{t-1}$ Muth (1961).

Can we propose a model for expectations that will extend the equivalence between distributed lagged expectations models and adaptive models, beyond the ARMA (1, 1) case? We showed that this equivalence can be achieved proceeding in two steps:

(a) Estimate the Box-Jenkins model of the univariate series $\mathrm{ARMA}(m, n)$ say $p(B)y_t = q(B)\varepsilon_t$ where p(B) and q(B) are polynomials in the lag operator B of orders m and n
(b) The most efficient, one step ahead, causal predictor for time t will in fact satisfy: $\hat{y}_t(1) = \arg\min E(\hat{y}_{t+1} - \hat{y}_t(1))^2 = E(y_{t+1}/y_t, y_{t-1}, \ldots, y_1)$ with E = expected value operator, can be derived from the estimated Box-Jenkins model as

$$\hat{y}_t(1) = -\hat{p}_1 y_t - \cdots - \hat{p}_n y_{t-n} + \hat{q}_1 \hat{\varepsilon}_t + \cdots + \hat{q}_m \hat{\varepsilon}_{t-m}$$

This is equivalent to the self-tuner predictor, Wittenmark (1974), used in identification of linear models in Control Theory, Hernández et al. (1979) and provides a general way to model adaptive expectations.

3.5 *Granger's Causality, RE and AEx Equivalence*

Two surprising results were observed when using time series analysis (prediction without theory). One was the good forecasting performance of univariate models, Nelson (1972) compared with structural econometric models. A reasonable explanation of this fact was that structural econometric models contained specification errors and that one should use small models with rational expectations. We preferred the alternative explanation that fast and frugal expectations, as adaptive expectations (AEx) could approach well theoretical rational expectations, RE. Aggregated people expectations, could manifest collective intelligence after all.

The second one was the absence of Granger's causality (GC) among variables such as money and prices that were related according to theory. This result reinforced our alternative explanation. We interpreted these results as an unexpected welcome proof that under stable regimes, adaptive (prediction without theory) and rational expectations (prediction according to theory) may be equivalent. Rational agents may draw on an information set larger than just the history of the variable being forecasted, including the structure of the relevant system describing the Economy. But if there was no Granger's causality this larger information set does not improve prediction and AEx are equal to RE.

Under what conditions should this be true?

According to Granger (1969), a variable y is not caused by the variable x if $\hat{y}_t(k) = E(y_{t+k}/A_t) = \arg\min E(\varepsilon_t(k)/A_t)^2 = E(y_{t+k}/A_t - x_t) = \arg\min E\ (\varepsilon_t(k)/A_t - x_t)^2$, where A_t represents all available information at time t and $A_t - x_t$ excludes this information about x at the same instant t. Therefore, if y is 'caused' by x in the sense of Granger, a pure extrapolative model of expectations about y is irrational in the sense of Muth, and y is endogenous in any model which includes *x*. This is the relationship between Granger's causality, rationality and endogeneity. If on the contrary there is no GC, extrapolative and the corresponding adaptive prediction are equivalent to rational expectations.

Of course, for a given sample a variable *x* may not "cause" in the sense of Granger, another variable *y*, even though according to theory *y* is influenced by *x*. As we reiterate along this chapter Economic Theory is contextually dependent. For example in the case of monthly money and prices time series of the Spanish economy in the period 1965–1976 of great stability, causality was not detected. However, for the German hyperinflation period 1920–1923 GC was detected. For the complete analytical details about when this may occur, see Hernández-Iglesias and Hernández-Iglesias (1981).

For our purpose here, the important message is that the non-GC results indicate that equivalence between extrapolative and rational expectations occurs frequently for relevant economic variables in stable periods.

"***If this is so, there were strong implications for macroeconomic modelling. The observed coincidence of efficient extrapolative expectations and rational expectations is more the rule than an exception for most available historical samples***" (Hernández-Iglesias and Hernández-Iglesias 1981). Ecological knowledge; simple rules, will be equivalent to constructive knowledge according to fully rational rules. In stability periods, RA and AEx are equivalent and citizens do not have to behave as econometricians to form efficient expectations.

4 What Can We Learn from the 2008 Crisis?

Many developed countries are still suffering the crisis that started in 2008. Others have partially recovered at a cost of low rate of growth, a great increased in public debt and more inequality. Great debt, quantitative easing and very low interest rates led to liquidity trap and little options were left for policy measures to increase investment and demand to recover growth. The crisis has shown that orthodox economist pay attention to the beauty of their models, but they forget that Economic Theories are contextual and Economics in the first place has to deal with institutional design. Real state capital with finance capital and insurance are at the heart of current misunderstandings of the economic crisis and recession. Financial and real state credit (FIRE) are not in the econometric models. An urgent task is to include models for the FIRE sector and the corresponding prudential rules. It is a task of Political Economy and Institutional Economics, not of Economic Policy.

The crisis is overlapping with ongoing changes in the supply side: Radical changes in production, coming from the second generation of Information and Communication Technologies. All along the second half of the last century, the effect of technological progress has been non-disruptive and smooth. The technological changes nowadays, are affecting production, management practices, and the labor market in a disruptive way. This means that we need proper answers to the questions raised by the actual crisis and by the ongoing changes in the supply side of the Economy.

4.1 Few Economists Saw the Crisis Coming

Everything was going well. As in the years before the seventies crisis it was a time of great moderation. Macroeconomists were proud of their models and unconscious of a major pitfall: They forgot to include a FIRE module and its relationship with both the monetary and the supply module. In 2008 the perfect storm was building up, the Minsky moment arrived and the self-assured hubris among economists was shaken.

Great economists and policy makers did not saw the crisis. I choose a sample of quotations. Ben Bernanke at the meeting of the Eastern of the Eastern Economic Association 2004. "*One of the most striking features of the economic landscape over the past twenty years or so has been a substantial decline in macroeconomic volatility... Several writers on the topic have dubbed this remarkable decline in the variability of both output and inflation "the Great Moderation." Similar declines in the volatility of output and inflation occurred at about the same time in other major industrial countries, with the recent exception of Japan, a country that has faced a distinctive set of economic problems in the past decade*". Japan was a good warning that something could be wrong but it was better not to spoil the happy days.

In an October 12, 2005 speech to the National Association for Business Economics, the then Federal Reserve Chairman Alan Greenspan spoke about the "*development of financial products, such as asset-backed securities, collateral loan obligations, and credit default swaps, that facilitate the dispersion of risk... These increasingly complex financial instruments have contributed to the development of a far more flexible, efficient, and hence resilient financial system than the one that existed just a quarter-century ago.*" Greenspan had in February 2005 asserted the US House Financial Services Committee that "*I don't expect that we will run into anything resembling a collapsing [housing] bubble, though it is conceivable that we will get some reduction in overall prices as we've had in the past, but that is not a particular problem.*"

In August 2008 Blanchard (Nobel Prize 2012), claimed, "*For a long while after the explosion of macroeconomics in the 1970s, the field looked like a battlefield. Over time however, largely because facts do not go away, a largely shared vision both of fluctuations and of methodology has emerged. Not everything is fine. Like all revolutions, this one has come with the destruction of some knowledge, and suffers from extremism and herding. None of this deadly however.* ***The state of macro is good***"

Nevertheless some economists like Shiller, saw the crisis coming, but in general those who saw it, are unconventional economists such as Keen, Minsky and Hudson. The crisis was there but econometricians did not care.

4.2 The Crisis as a Classical Financial Panic in a New Financial System

One of the ideas I insist on in this chapter is that Economics must refer to a specific context that is only certain in the present and in the past. The first thing a reasonable economist can do to understand the crisis is to look for historical similar cases. Bernanke (2013) makes use of the 1907 financial panic episode caused by failures in the financial system of the time to analyse the failures in the actual financial system and the agenda to repair them. I prefer to make use of the last Japanese crisis because starting 20 years earlier can give us not only information about the crisis but also about the problems faced to recover growth, once confidence in the financial systems has been restored.

Since 1961, the Bank of Japan (BOJ) attempted to control directly the volume of commercial bank credit by providing lending targets for selected banks. This policy was applied to a subset of lending institutions. This led to low interest mortgage loans with higher risk. Low interest rates and loose monetary policy fuelled a strong growth and raised stocks and housing prices. Following the Plaza Agreement in 1985, the yen appreciated from around 240 yen to the USD to about 120 yen in less than a year. In response, the Bank of Japan lowered interest rates from 5.5% down to 2.5% in 1987. This dramatic easing of monetary policy at a time of economic strength provoked an explosion of real-estate transactions and high stock prices.

In 1988, Prime Minister Nakasone reduced corporate tax rates from 42 to 30% and reduced top marginal income tax rates from 70 to 40%. The combination of easy monetary policy and expansive fiscal policy led inevitably to the Japanese 1990 stock market crash and a deep fall of housing prices. Equity and asset prices fell, leaving overly leveraged Japanese banks and insurance companies with books full of bad debt. The financial institutions were bailed out through capital injections from the government, loans and cheap credit from the BOJ, and the ability to postpone the recognition of losses, ultimately turning them into "zombie banks" that in turn kept financing "zombie firms" for political interests. Housing prices (index 100 in 1975) fall from 215 (1991) to 130 (1994) and to 100 (2003). A loss of 60% in housing wealth over the crisis period.

A strong increase in government spending and stagnant revenues raised the government debt to more than twice the GDP and a fall of GDP from 7, 1% in 1988 to −5, 5% in 2009 and to 1, 6% in 2013. The lesson is clear. Intervention of the BOJ trying to correct the yuan appreciation, stimulating private investment by monetary expansion, cutting interest rates and fiscal expansionist polices to increase consumption, led to the bubble. Monetary policy after the crash was limited because a liquidity trap and fiscal expansion was not possible due to the huge public and private debt from zombie banks and their linked corporation's debt. As a last resource, the BOJ was accepting the bad practice of buying debt in private hands. The recovery is being slow and poor. The cause of the crisis was not a failure of the financial system but a misuse of Economic Policy forcing the BOJ to accept too low capital ratios of the banks.

Blanchard's Macroeconomics textbook (2013 6th edition Chap. 9) provides a detailed account of the causes and policy responses of the 2008 crisis. I summarize the main facts with personal comments. There are some differences with the Japanese crisis. The trigger of the crisis was a fall in the stock prices in Japan, whereas in the US crisis it was a fall in housing prices since 2007. The difference is not relevant since both end up correlating after a short time. Japan was more exposed in relative terms, because the big pension funds and foreign speculative investment.

The other difference is more important. In the 2008 crisis, a complex "engineering" (an "Alphabet Soup" is Blanchards's label) of products and financial agents (SIVs) not limited to banks and insurance companies were created to avoid the bank's capital ratio limitation to lending. This is a "shadow financial system" where risk was difficult to assess, opaque, fuzzy and without traceability, Fig. 10. Investors accepting Market-Based securities (MBS) and other forms of securitization were acting without been aware of it as "bankers" in the darkness. These led to at least the following problems, Blanchard (2009):

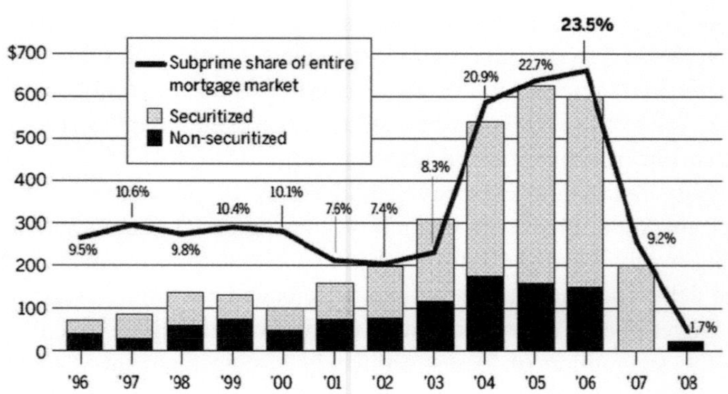

Fig. 10 The shadow finance. *Source* National Commission on the Causes of the Financial and Economic Crisis in the United States (2011)

- Assets created, bought and sold were riskier than they appear. With expectations of rising housing prices, subprime mortgages risk was acceptable. However, if housing prices fall, at some point many mortgages would exceed the value of the house, leading to default and foreclosures.
- Securitization makes difficult to value assets on the balance sheets of financial institutions. This product was in the market in the nineties, but in 2008 more than 60% of the US mortgages were securitized and income streams from these securities were trenched to offer investors different risk flows alternatives.
- Securitization and globalization led to increasing interconnection of the financial institutions. However, this interconnection destroyed the theoretical advantage of risk avoiding by pooling mortgages. The variance of the sum of m independent random variables with the same distribution is m times less than the variance of a single one, but the tranches of the pooled mortgages are not independent. The financial system became anything but resilient as Greenspan thought and it favoured free riding and moral-hazard behaviors.

The first aid to cope with the panic was restoring confidence, allowing banks and other financial institutions to borrow from the central banks, government bailed outs to banks, insurance companies and corporations, cleaning up banks separating the toxic products (the Troubled Asset Relief Program, TARP) etc. When the financial storm ended, its destructive effects were evident. For example, in the period, 2007–2015 Spanish families have lost 30% of their wealth. However, this reduction comes from their housing wealth. Their financial assets wealth has increased. That means that saving during the storm was the family's response. This in turn depressed consumption and demand. To put the Economy back in track a further fiscal push was needed since monetary policy, as in Japan, was blocked by a liquidity trap. The result in Spain is an increase in public debt of 60% and a rise in unemployment to 22%. What is much worse, the inequality gap has widened.

The cause of the 2008 crisis, unlike the Japaness one, was changes in the financial system. The great moderation had fooled not only macroeconomists. Financial institutions and regulators also underestimated the financial risks. The result was a financial structure exposed to free riding and moral hazard. Any reform of the sick financial system has to minimize these two lurking dangers.

4.3 A New Financial System. Creating a Socially Useful Financial System

We agree with many economists worldwide that the post crisis task needs more of structural reforms of the financial system and less of Economic Policy firing, although they should go in parallel to recover growth. "*But it's not going to come easily from a political point of view. We need to make courageous decisions, which we've been talking about for a long time*" (Legarde 2016). To minimize the two lurking dangers, free riding and moral hazard, we need structural reforms and prudential rules.

Looking to the Japanese post crisis one cannot avoid the general feeling that something more fundamental is missing if not wrong in Economics. This feeling has led to the creation of new institutes and scientific societies that joined those already rethinking Economics. One of these is the Institute for New Economic Thinking (INET) founded and financed by Soros. When I was preparing my Keynote session for the Artificial Economics Conference in Barcelona (2104) I found a talk by Bezemer at the 2012 INET meeting in Berlin titled, "*creating a socially useful financial system*".

In the talk, he separates the credit that goes to the "productive" system from the credit that goes to the financial and real estate sector (FIRE). In today's financial world most credit is not spent in creating added productive capacity but to buy assets already issued. According to Bezemer, about 80% in the English-speaking countries bank loans are real estate mortgages and much of the balance is lent against bonds already issued and stocks. To accomplish structural reforms, we really need economic and political institutional arrangements to develop coherent alternatives to mainstream analysis. Why? Because mainstream Economics and consequently the corresponding econometric models do not include the credit-debit balance in the financial investment, insurance and the real state sector (FIRE). Incorporating the FIRE sector in financial models is going to be a difficult battle because this will imply to unveil the effects of the "rentier sectors" on the Economy and to change macroeconomic financial accounting. Finance is not in the mainstream version of Economics.

The accounting approach of finance and credit has its roots in a monetary view of the circular flow of the Economy whose representatives are Marx, Schumpeter, Kalecki, Minsky, Godley, Baker, Keen and Hudson and at times Tobin. It is an heterodox position with respect to the classical and neoclassical economics that maintained that in the long run money is neutral, ignoring that money and credit are not the same thing since "financial engineering" and shadow banking arose. This crisis is an empirical confirmation that money and credit matter.

What happens around 1985 for the first time in the history of the Economy? Figure 11, from Bezemer's talk, shows credit to the dual sectors, FIRE and non-financial, as a percentage of GDP. The ratio of loans to nominal GDP for the FIRE sector, increased from 50% in 1985 to 250% at the beginning of the crisis. The effect is clear: A tremendous inefficiency in the financial system that is necessary for maintaining the economic activity. The storm began to build around this year.

I included, of course, this graphic in my Keynote as one of the challenges to Economics: How to model a socially valuable financial system that includes money and credit. I considered, at the time, his social accounting approach a seminal contribution to understand the crisis and the necessary reforms and much more. I was right. I refer some of Bezemer's messages since then:

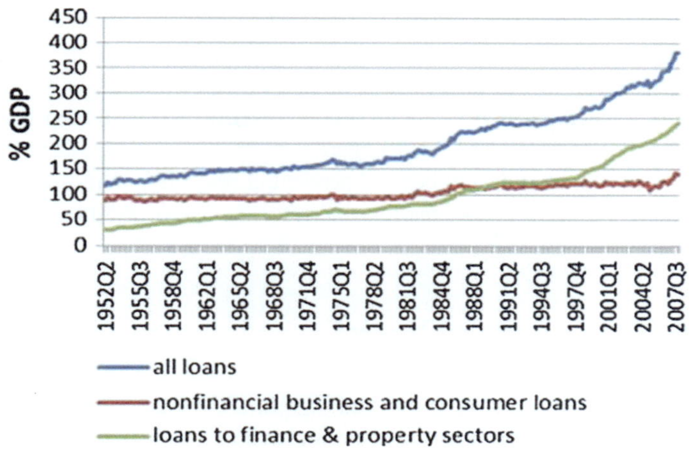

Fig. 11 Lending to the real and FIRE sectors. *Source* Bezemer (2012b)

(i) The credit nature of money has macroeconomic significance, Bezemer (2012a)
(ii) The FIRE sector pumped wealth in the form of revenues from the real sector, Bezemer (2012b, 2016)
(iii) Banks' lending does the necessary job, creating enough money (through the money multiplier) to meet the demand of the productive sector for goods and services and a little more, to avoid occasional illiquidity and to stimulate aggregate consumers demand. Schumpeter was right. However, credit to the FIRE sector hinders growth and distorts wealth distribution. A functional classification of credit is due, Bezemer (2014)
(iv) Macroeconomic models have to introduce a money and credit sub model making explicit the overall model of the FIRE sector. Hudson and Bezemer (2012)
(v) Finance is not the Economy. Rent is not Income. FIRE credit is unproductive credit and imposes overhead costs to the production sector, Bezemer and Hudson (2016)
(vi) Credit flows to non-financial business have a positive effect on growth but credit flows for mortgages and other assets have no effect or even have a negative effect, Bezemer et al. (2016)
(vii) Bank credit to real estate and financial asset markets, increases income inequality. Credit to non-financial business and for household consumption supports broader income formation, decreasing income inequality. Since the nineties has been a progressive shift in credit towards the FIRE sector and consequently increasing the inequality gap, Bezemer and Samarina (2016)
(viii) Can ABM models help developing socially useful macroeconomic modelling? Bezemer (2012b), Schasfoort et al. (2016) and Bezemer et al. (2016).

We have learned that the failure of considering credit and money in its duality (real productive credit and FIRE credit) is the core cause of the crisis. This time the cure and the recovery of the crisis will need not only an explanation but also a new design of the financial system to avoid a new crisis, which will deepen inequality and undermine democracy. Yes, Mrs. Legard is right. We need courage because it is a question of institutional design beyond prudential rules. It is a question of politics and economics. A return to the institutional reforms task of Political Economy.

5 Economics as a Social Science

Economics deals with social behavior and therefore inherits complexity. Trying to be a science, it is based on theories according with empirical evidence. However since the society and economic institutions change, the Economy evolves and standing theories in Economics have to be either completed or replaced by new theories to accommodate new contexts. In this chapter, we have referred to three challenging contexts within a period of 60 years. The interval of time between two different contexts that force to change or to complete the Economic relations is very long. During these periods of "great moderation" economist can enjoy their wisdom and politicians are happy applying the accepted rules. One could expect that the economists, being aware of the contextual validity of their theories, would tolerate methodological diversity. It is not like this.

5.1 What Is Wrong with Economics?

The unrealism of its assumptions. The economist build models assuming things which don't occur in the real world, as long as their models follow from mathematical reasoning and have predictive capacity.

Equilibrium thinking. Equilibrium conditions are the closure of their models, perhaps because when non-narrative Economics started, the world was fascinated with physical laws. Economist aimed to find universal laws equally compelling in the Economy. I remember using the model of a magnetic field to explain market equilibrium; the flow of oil through a pipeline in terms of hydraulic horsepower and the internal pipe diameter to justify a Cobb-Douglas production function or showing them that the funny parameter called in maths the Lagrange's multiplier was just a shadow price. I may be excused because I was teaching Economics to Engineering students. I agree that Physics has a lot to contribute to Economics today, but Economics is not like Physics. In Economics, there are no general laws. Laws are contextually valid.

Methodological individualism. Individuals, their choices and decisions are the sole units of analysis, without much reference to Psychology or Sociology. Tags and kinship do not count. The agents are fully rational, equally motivated, something that Keynes criticized with his metaphor of the beauty contest. Of course, game theory has relaxed somehow this assumption, but this autism with respect other sisters disciplines, has caused very bad effects in financial markets.

Formalism. All the accepted economic laws ought to be formalized in a mathematically coherent form. The problem with this formalism is that "*the great virtue of maths is that it does make precise things which ought to be precise. Its great defect is that it imparts a false precision to things which cannot and ought not to be made precise*" (Skidelsky 2015). Many of the results in EE and Artificial Economics (AE) can be described in a narrative or graphical way and of course, when it is possible, in a mathematical formalism. You can proof by numerical calculus where the maxima of a function is, or use your computer to graph the whole function and select all the local or global maxima. You can try to find out the probability distribution of say sin x, where x is a random variable that follows a Poisson distribution or just simulate and graph the results. My choices will be in both cases the second alternative.

The filter of ideological context. The very name of Political Economy emphasizes the dependence of the Economy of the political and ideology context. Many good contributions in Economics are left aside because they are not in agreement with the prevailing ideology.

Academic dictatorship and monoculture. The INET-CORE project is a response to the growing student's protests against Economics teaching nowadays. Here is Tirole's 2012 Nobel Prize Winner writing a letter to the French Secretary of State in Higher Education and Research and the AFEP response.

"*...May I inform you of my concern about a continuing rumor about the creation of a new section of the National Council of Universities named «Institutions, Economy, Territory and Society». Should this rumor be confirmed, it would cause a disaster for the visibility and the future of research in economics in our country. It is especially important for the community of academic teachers-researchers to be endowed with a single scientific assessment standard, based on the ranking of the journals of the discipline and on an external assessment by internationally prominent peers. It seems inconceivable to me that France would recognise two communities within the same discipline.... Self-proclaimed 'heterodox' economists have to comply with the fundamental principles of science.*"

Response of the heterodox Association Française d'Economie Politique (AFEP)

1. *"The claim that «heterodox» economists want to escape the assessment of their research is preposterous. Our proposition of an authentic peer evaluation of research, based on a variety of publications, is different from the current ranking of journals and the perverse quantitative bibliometric norms of assessment that it implies."*
2. *"The proposal to create a commission with Nobel Prize and Clark medals winners is similar to take a representative sample of a Papal conclave to decide about the legitimacy of a demand by a minority of protestants."*

The position here is that Economics is a science, we agree, and there is only one-way to do science, we disagree. Economics is a social science with complexity far beyond a "pure" science. Tolerance and academic diversity is necessary. 50 years for Experimental Economics to gain acceptance in mainstream economics give us a measure of the orthodox economist's intolerance.

5.2 *Mathematics and Game Theory Cannot Deal with Social Complexity*

Complexity is a term used in many ways according to different schools in Economics. For our purpose, we mention three types. A dynamical system is complex if it endogenously does not tend asymptotically to a fixed point, a limit cycle, or an "explosion". Alternatively, a situation exhibits complexity when there is an extreme difficulty of calculating solutions to optimisation problems. Another source of complexity appears when it is required to deal with agents with bounded rationality yet strategic behavior and these heterogeneous agents should learn from the decisions of others. This is where game theory has attempted to extend market design from a constructivist mathematical approach with limited success.

Let me illustrate the last source of complexity with my preferred persuasive argument against forcing Economics into the Procrustean bed, and the need of new methods in Economics: The generative method of Experimental Economics (EE) and Agent Based Modelling (ABM) applied to Economics, that is, Artificial Economics (AE). The example is taken and extended, from the first edition of Pindyck and Rubinfeld (1995) Microeconomics textbook.

Three contestants, A, B and C, have a balloon and a pistol each. From fixed positions, they fire at each other's balloon. When a shot hits the balloon and breaks it, its owner is out of the game. When only a balloon remains, his owner is the winner and receives a $1000 prize. At the outset, the players decide by lot the order in which they will fire, and each player can choose any remaining balloon as his target. Everyone knows that A is the best shooter and always hits the target; that B hits the target with probability 0.9 and C with probability 0.8. Which contestant has the highest probability of winning the $1000? When I asked my students to advance an answer within 5 min, some will come up with a reasonable and correct one: Contestant C.

What can we learn from this example? An intelligent and knowledgeable student will use probability calculus to work out the right solution, should he have enough time: procedural rationality, in terms of Simon (1982). This was not the case given the time available to answer. How they arrived to the correct answer? Because "as in life the success is for the mediocre" or because "whenever the balloon A is not broken, B or C will shoot to A". Both answers come from "social knowledge" not from "constructive knowledge".

When dealing with markets, should one restrict to full rational agents or to agents that use fast and frugal decision rules? Posada and López-Paredes (2008).

Individual constructivist methodology does not lead to better models of learning decision than ecological and social knowledge.

Let us extend the example. Once we know that the most likely winner is C, let us repeat the game in such a way that we can keep enjoying the betting game, and the shooters earn their salary. Who will be now the winner? Mathematics will not provide an answer and game theory will be useless for such a simple problem. However, the answer is there: Experiment! See if there is a similar situation in Economics (natural experiment). Yes there is. The dominant firm model or "leave and lets others live" model of oligopoly can provide an answer. There is no need to trespass the constructive mathematical approach, of Industrial Organization textbooks.

However, is the dominant firm model as explained within a constructivist approach correct? You may try to justify that is a correct model with empirical econometric tests, but this is not a sufficient consistency test. What do I mean? Am I saying that the constructivist models and the corresponding econometric test are not enough to certify the scientific validity of economic models? Yes, that is what I mean. One needs to complete the model doing a previous check of internal consistency. Identifiability in terms of the Cowles Commission, or in today's terms, explaining the dynamics to reach the equilibrium model that you are rescuing from statistical data as Smith, wisely warns us:

"As I see it, there is no rationally constructed science of scientific method. The attempt to do it has led to important insights and understanding, and has been a valuable exercise. But all construction must ultimately pass ecological or 'fitness' tests based on the totality of our experience" (Smith 2008).

5.3 *Experimental Economics*

EE is an approach to Economics that uses human agents in the classroom, the computing laboratories or in the field, as an engine for generating substantive propositions, learning about behavioral assumption or testing and calibrating alternative economic models.

It has developed from disagreements of the economists with some of the strong assumptions of orthodox Economics, replicating observed individual and social behavior. It is the result of the integration of the Economy with Psychology and Social aggregate behavior. Today the integration of Economics with Psychology has been achieved through EE. The integration with Social behavior is pending. We regret the change in title of the Journal of Socio-Economics where Smith published a revised version of his N. Prize speech to the new title: The Journal of Behavioral and Experimental Economics, forsaking the term social. A survey in one document of the current state of EE for those unfamiliar with it is unusual given his wide range of applications. Nevertheless Chakravarty et al. (2011) can be appropriate.

EE allows introducing in the economic models, agents with bounded rationality, heterogeneity and social behavior. We agree with Davis (2016) "*The basic rationale for the experimental method has remained pretty much the same over the years: Using the lab can overcome the difficulties of finding data in many naturally occurring settings; the lab has a high degree of "internal validity" because it allows the researcher to change one variable at a time. All of these advantages have contributed to its acceptance as a standard tool in the economist's toolkit*". This "internal validity" is the fitness test quoted above from Smith.

Following the Nobel Prize award in Economics to Smith and Kahenman in 2002, the use of EE increased. The Nobel Prizes to Ostrom in 2007 and to Roth in 2012 certified that EE is one of the pillars of Economic Methodology. The progress has been very important in the main three streams: Market design, games, and individual or social choice experiments. In what follows, we will deal only with impersonal experiments.

Nowadays there are two windows to look at the Economy.

The *value window*. A top down constructivist approach, with two generations. 1st g: Market equilibrium with homogeneous and fully rational agents. 2nd g: Game theory and personal exchange, with heterogeneous agents and endogenous growth (Fig. 12).

The *exchange window*. Ecological learning and social intelligence. Socially inspired methods in Economics and Management. 3rd generation: distributed

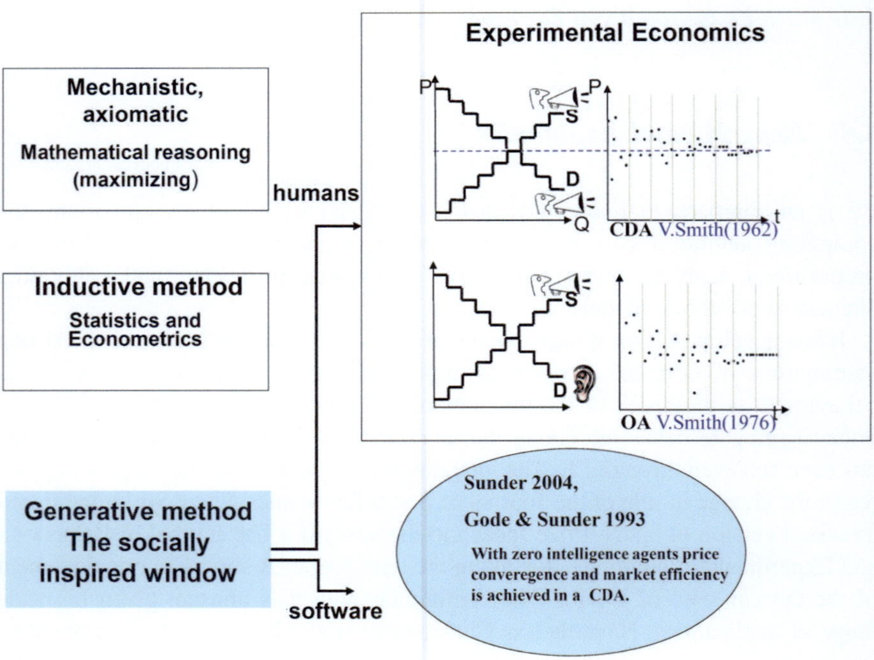

Fig. 12 The "exchange windows" for managerial and economics institutional design

intelligence with heterogeneous bounded rational agents and bottom up modelling. EE extended to soft agents: Artificial Economics (AE).

In one of the earliest market experiments Chamberlin (1948) conducted a classroom trading exercise that was designed as a test of the competitive model in conventional market theory. A group of students were buyers and received a card with a reserve value written on it, and the other group, were sellers and received a card with a reserve marginal cost written on it. Students then walked about the room, and buyers and sellers could negotiate over the terms of trade. When a deal was agreed, the price was written on the blackboard. This test produced trade volumes in excess of the competitive equilibrium price, and trade prices that were quite variable.

Smith (1962) redesigned the experiment with bidding rules similar to those used in equity markets. All bids and offers were centrally and publicly recorded instead of allowing the students to mix in the room and bargain over prices as in Chamberlin's experiment. This modification of the trading rules, which is known as Continuous Double Auction (CDA) since both buyers and sellers are active, leads quickly and accurately to the predictions of the competitive market model.

Let us modify the bidding rules: Buyers are listening to sellers' offers but do not make their bids public. What is the expected equilibrium price? Microeconomics has no answer, except at the expense of forcing an equivalence between information and randomness. Asymmetric information is measured by the effects on transaction costs, but cannot be included in the dynamics of the model. Of course, EE can, Smith (1976). Since information is costly, those who hide information should have an advantage and the equilibrium price is expected to be below the price of the symmetric auction, as it happened.

There are deep lessons from these seminal experiments.

(i) Collective intelligence. Experiments using the CDA institution converge reliably to the competitive price even with few participants, and neither the buyers nor the sellers need to have information about the values or costs of each one in the market. They are guided by sharing prices information towards the emerging equilibrium. They are standard, yet purposeful citizens. However, they exhibit collective intelligence. Markets as economizers of information: Examination of the "Hayek Hypothesis", Smith (1982). Hayek's (1945) argued that *price mechanism serves to share and synchronize local and personal knowledge, allowing society's members to achieve diverse complicated ends through a principle of spontaneous self-organization: that which results of human action but not of human design.* Spontaneous order.

(ii) For sociologists, social (ecological) learning is the process by means of which agent's acquisition of new information is caused or favoured by their being exposed to one another in a common environment, Conte (2002). The collective wisdom of Hume.

(iii) EE allows detailing the dynamical process, and it will serve us to the final consistency test to accept or reject a model, beyond the econometric tests.

(iv) The CDA allows us to experiment the dynamics of other well-known market models such as duopoly, monopoly and assets markets.
(v) The auction itself is a powerful "mathematical solver" for even zero intelligent agents that are absolutely aliens to the perfect microeconomic market model used to find the market equilibrium price/output pair. This is of great importance to Management Engineering. Auction inspired methods (*socially inspired methods*) can be applied to many management activities, such as forecasting, project and ideas selection, collective intelligence, marketing research, yield management and dynamic pricing under the name of "prediction markets" ("prediction auctions" should be a more accurate name) and crowdsourcing, Arrow et al. (2008). Many innovative as well as consulting firms are using already auction-inspired methods.
(vi) We can now interpret the conventional supply and demand model as a gadget, although useful, of what the market does. If we are interested in forecasting market responses to variations in the demand or supply the constructivist market mode is still a useful tool.

Consequences of points (i) and (ii). Closing the rational expectations debate. We conclude in the referred paper (Hernández-Iglesias and Hernández-Iglesias 1981) that, "…*they* (the analytical results) *indicate that equivalence between extrapolative and rational expectations occurs frequently for relevant economic variables in stable period*" We have now an EE support of this conclusion. The bizarre statement that expectations made by the citizens should be as rational as those of an econometrician is compatible with adaptive expectations, because a simple individual behavior might reveal ecological intelligence, to match the expectations of a good and lucky econometrician.

Consequences of point (iii). Testing economic models far beyond statistical tests. *The Ecological "fitness" test*. Going back to our balloons contest. The dominant firm model is a standard in Industrial Organization and it applies today in antitrust cases. Does it pass the ecological fitness test? No, it does not. "*The predicted market price and dominant firm output depend on arbitrary simplifying assumptions about behavior – in particular, that only the large firm is aware or perceives that it faces a downward-sloping demand. All other firms are presumed to behave as price-takers, reminiscent of the theoretical myths that a CE is driven by "price-taking" behavior. One implication of this assumption is that if the fringe supply costs are rotated using the equilibrium as a pivot so as to leave the dominant firm's equilibrium price and output unchanged, this will not affect anyone's behavior*" (Smith 2007).

Rassenti and Wilson (2004) studied the behavior in a dominant firm environment using two separate institutional market rules applied to the dominant and to the followers (fringe) firms. One was a posted-offer auction common in retail markets. The other a sealed offer auction. All the firms submit their sealed offers. The uniform clearing market price was determined by aggregating the offers and the actual demand. In the first case, the dominant firm quite often produces more than the dominant model and at higher prices. In the second case with a low elasticity of the aggregated followers supply, the dominant firm produces more than the dominant firm prediction for a wide spread of prices around the predicted price. We can use the dominant firm model of price leadership only as a benchmark of the true underlying competition. An example that **Industrial Organization needs Experimental Economics.**

5.4 Artificial Economics (AE)

In the late nineties, a young assistant lecturer, Adolfo López-Paredes, was developing an EE economics Lab (LABEXNET) as part of his Ph.D. in my Department. When helping him "shopping in the field" of EE, I came across a short version of a weird draft, that talked about intelligent-cognitive artificial agents and it reported results of simulating a duopoly. I could see there a reach set of results including the dynamics towards equilibrium. I was fascinated by what I was reading, signed by Professor Scott Moss at the Centre for Policy Modelling, Manchester. I wrote to S. Moss asking for more details and he sent me a report of the research undertaking at the CPM with a note "*good luck*". We arranged a visit of Adolfo to S. Moss and he was accepted to work at the CPM. With a degree in Industrial Engineering, not in Computing, he was able to get his hands on the SDML a very advance platform for social simulation. In 1999 he obtained his Ph.D. (López-Paredes 1999). Some of the findings were published in López-Paredes and Del Olmo (1998), López-Paredes (2000) Hernández and López-Paredes (1999, 2000) and the narrative content in López-Paredes et al. (2002). He used the EE results from the LABEXNET to initiate the AE simulations. EE and AE are complementary.

In the early nineteen's I created a group of young colleagues and Ph.D. students interested in the Socio-Economics Applications of Agent Based Modelling (InSisoc) at the Universities of Burgos and Valladolid. I participated as founder of ESSA and member of its Management Committee and later on of the Artificial Economics Conference (ACF). We hosted the ESSA conference in 2004 and the ACF in 2009.

Under the label of SSC there are joint conferences of the ESSA (European Social Simulation Association), representing Europe; PAAA (Pacific-Asian Association for Agent-based Approach in Social Systems Sciences), representing Asia and Oceania; CSSSA (Computational Social Science Society of America), representing the Americas. Each year there is a Summer School in Social Simulation for young people. This short history shows that AE is growing rapidly with young researchers in any field of Economics and Management.

AE economics is what it does. In the JASSS journal there are papers covering AE and social simulation. In the 12 AE annual conferences proceedings: http://www.artificial-economics.org/, there are more than 200 papers which cover a wide range of research in Economics, Finance and Management and show how AE has evolved in the last decade.

Defining AE is controversial because it is at the intersection of Artificial Intelligence and EE. Since the product is always a computer programme, Agent Based Computational Economics (ACE) is being used as an alternative. The heritance of ACE from Dynamics and Control leads to a bias towards computational properties of mathematical top down models, the *value window*, against the generative bottom up models of EE and Agent Based Modellig (ABM), the *exchange window* (market oriented). For more on this issue see Shu (2012) and the keynote session at the AE conference 2015, (Izquierdo and Izquierdo 2015). They recently updated a brief but informative view of AE in https://en.wikipedia.org/wiki/Artificial_economics.

AE is EE with soft agents. In AE, agents are no objects. Objects do it for free, agents in AE do it for "money". They are autonomous and purposeful. They can be heterogeneous, and exposed to another agents in a given environment. The agents can be endorsed with differences in information, learning capacity and decision choices. The modelling process is bottom up, generative. AE inherits the entire EE's legacy. In particular, as in EE, economic models exhibit the dynamic process towards equilibrium and how the models adapt to external or internal changes. Been a generative process it does not impose equilibrium conditions.

"*…it will be clear now that the main rationale to do AE is that it expands the set of assumptions that we can explore. The reason is that the set of assumptions that we can investigate using computer simulation is not limited by the strong restrictions that mathematical tractability imposes, so a whole new universe of possibilities opens up. This point is particularly important in the study of socioeconomic processes, which–due to its complex nature–are oftentimes difficult or impossible to address adequately using a purely deductive approach only. The theoretical analysis often requires so many simplifications to ensure tractability that the correspondence between the real world and the model assumptions ends up being disappointingly weak. Thus, using the AE approach we have the potential to understand socioeconomic processes better, and al-so to assess the impact and significance of the simplifications made by the theoretical approach*" (Izquierdo and Izquierdo 2015).

The short definition of AE, encapsulates its limits and strength. A large number of agent decision-making models can be found in the literature, each inspired by different aims and research questions (Balke and Gilbert 2014). When we construct AE models, the agents must inherit bounded rationality as in EE. Dealing with aggregated impersonal models, as in Macroeconomics, the agents decisions should be fast and frugal and learning should be ecological. Therefore, it is necessary to avoid the excesses of endorsing the agents with constructivist capacity. Of course, one can use constructivist soft agents to check the compatibility under appropriated institutions with fast and frugal rules, testing again the Hume- Hayek hypothesis. Patching econometric models, such as the DSGE, with artificial agents with decision capacity far from common sense reasoning, is a bad practice. However, it seems unavoidable. Perhaps the on-going project Agent-Based Macroeconomics will contribute to clear this issue, Dilaver and Gilbert (2016).

5.5 Building AE: The Exchange Window

Following the legacy of EE (Smith 1989), to generate an AE model, we consider three dimensions that are essential in the design of any market experiment: The Institution (I) (it is both the exchange rules and the way the contracts are closed, and the information network), the Environment (E) (agent endowments and values, resources, knowledge) (A). They are represented in Fig. 13.

By mapping different arrangements of the elements of this triplet (I × E × A) into observed and forecasted outcomes (O), a host of experimental results can be obtained. This is quite important. You can try to specify the triplet of conventional microeconomic models and you will see that it is not a trivial task.

We must go beyond conventional EE if we want to control the agents' behavior (A) dimension of our experiments. We have to move from human to artificial agents as argued in Hernández and López-Paredes (1999, 2000), López-Paredes et al. (2002) and Posada et al. (2006a). Taking this step, a rich program of research comes up, just widening the many relevant findings of EE with human agents, and checking their robustness against alternative controllable agents' behavior.

The first experiment with programmed agents (Gode and Sunder 1993) was a big surprise. They confirmed that institutions matter. To the extreme that in a CDA price convergence and allocative efficiency was achieved, even with zero intelligent (poorly instructed but perceptive) agents. That spontaneous order arises in the CDA, thus confirming F. Hayek and A. Smith conjectures.

The Institution (I) may be, for example, the CDA of the Smith experiment commented above. The CDA market is the dominant institution for the real-world trading of equities, energy, derivatives, emissions permits, etc. The CDA imposes no restrictions on the sequencing of messages. Any trader can send a message at any time during the trading period. We consider a CDA with a bid-ask spread reduction. The only restriction to accelerate convergence is that a new bid/ask has to provide better terms than previous outstanding bids/asks.

The environment (E). Each trader is either a seller or a buyer. Each agent is endowed with a finite number of units. Seller i has ni units to trade and he has a

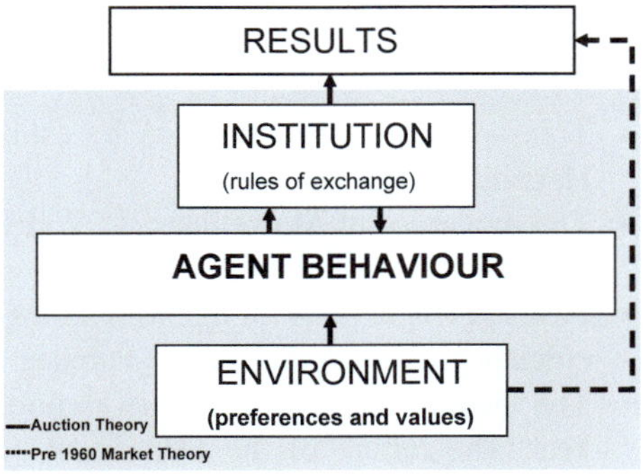

Fig. 13 Dimensions in the design of any market experiment

vector of marginal costs (MaCi1, MaCi2… MaCini) for the corresponding units. Here MaCi1 is the marginal cost to seller i of the first unit; MaCi2 is the cost of the second unit, and so on. Buyer j has nj units to trade and he has a vector of reserve prices (RPj1, RPj2… RPjmj) for the corresponding units. Here RPj1 is the reserve price to seller I of the first unit, RPj2 is the reserve price of the second unit, and so on.

The model may be restricted to homogeneous populations or to symmetric environments. It may have no environmental restrictions or any environment in terms of the number of traders, their units and the valuations of each trader. We can alternatively define market environments, both with symmetric or asymmetric supply and demand curves. We consider that an environment is symmetric if the supply and demand curves have opposite signs but equal magnitudes. Otherwise, we consider that the environment is asymmetric. The extreme case is when the supply curve and/or the demand curve are perfectly elastic.

Of course, should it be relevant, we may consider an external physical environment and the bargaining among the stakeholders or the auction will take place at a different level; the agents can share information from the physical landscape and incorporate it into the agent's environment. For a water management planning of a metropolitan area, Fig. 14, shows the model developed in López-Paredes et al. (2005) and Galán et al. (2009).

The agents (stakeholders) will negotiate to achieve a solution, bargaining about the global proposals or competing through an auction to select the winner. Of course once there is a winner proposal there could be fine-tuning "over the counter" agreements.

Agents' behavior (A). In a CDA markets, traders face three non-trivial decisions, Chen (2000): How much should they bid or ask for their own tokens? When should they place a bid or an ask? When should they accept an outstanding order of some other trader? We may define agents with different learning skills. The agents can choose their strategies in an evolutionary way according to individual and social learning for each trading period.

See the scheme in Fig. 15 for an ABM with different agents' learning endorsements.

Following this design procedure, research on impersonal exchange at InSiSoc have tested the robustness of the CDA with ABM simulation under very different scenarios and found quite relevant results in terms of auction's performance:

- Transaction cost in CDA can be estimated from the experiment (Posada and Hernández 2010)
- The dynamics of Marsallian and Walrasian instability in a CDA (Posada et al. 2008)
- Fast and frugal rules for agents learning maintain convergence, performance and efficiency (Posada and López-Paredes 2008)
- The institutional design is very important to achieve the objective of the markets: The failure of the EPA is an outstanding example of improper design (Posada et al. 2007)

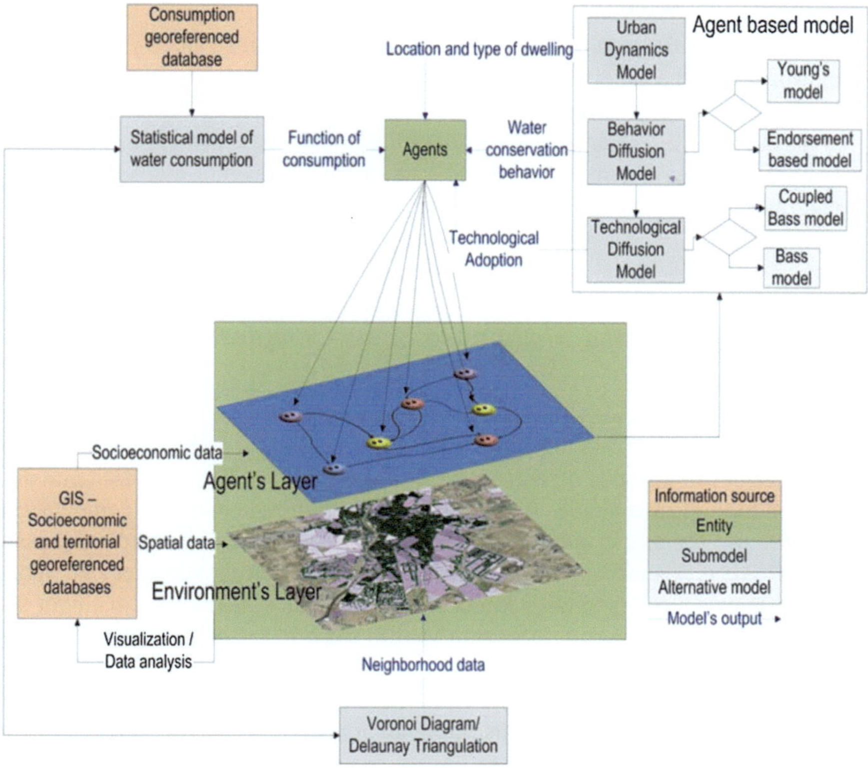

Fig. 14 SIGAME project architecture by layers

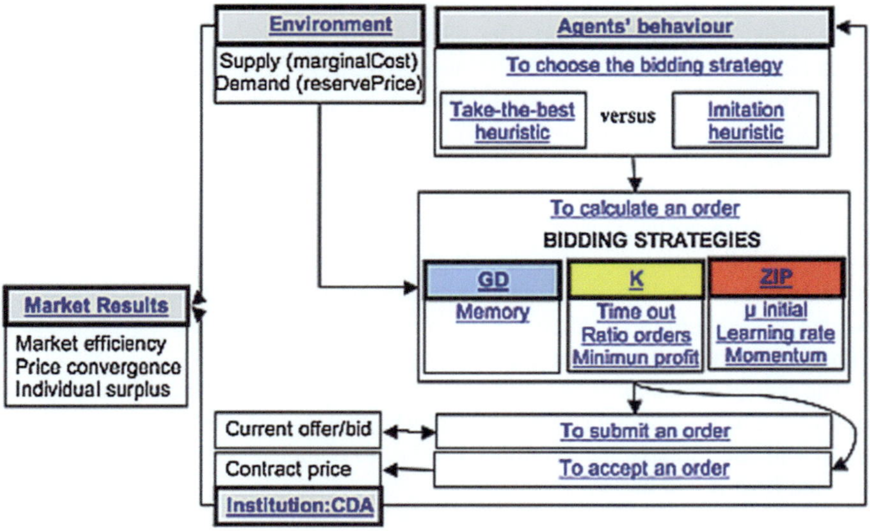

Fig. 15 Auction dimensions. Case of a CDA with 3 types of learning agents: SD, K, and ZIP

- Agent's intelligence and strategy matters. If we allow for agents with different learning capacities and skills, efficiency and performance can be achieved, but the surplus of the different groups of agents may differ. If the proportion of one kind of agents goes over certain threshold, then efficiency decreases (Posada et al. 2006a)
- Even for non-intelligent, yet perceptive and motivated agents, convergence to an equilibrium price and efficiency will be achieved (Posada et al. 2006b).

5.6 *Socially Inspired Methods to Solve Complex Problems*

We have seen that auctions (market exchange) are solvers for scarcity and choice complex problems, because they are an n-p hard problem or because an agreement has to be achieved introducing competence. In n-p hard problems, we use heuristics inspired in biological analogies such as genetic algorithms and swarm computation. Why not to use socially inspired methods in complex engineering and management problems? AE is Agent Based Modelling in the Economics or Management. When we use AE exchange methods to modelling physical systems, we maintain the original label as Multi-agent Systems (MAS), since it refers to a wider field of applications. In MAS, we try to build an artificial social simulation model to allocate scarce resources with alternatives uses, endorsing the model agents with behavior and motivations used in auctions models, given the self-regulating capacity of markets. In the MAS approach, we use a market metaphor to solve problems of engineering and management.

This approach has been used to solve many Management relevant problems. Landing and take-off schedules at airports (Rassenti et al. 1982) and SESAR (2014); optimization of freight transport (Bertsekas 1990); management of sections of the rail network (Parkes and Unga 2001); to develop flow shop production programs, Wellman et al. (2001); management resources in a portfolio of projects etc.

To illustrate the idea, we will briefly describe the MAS application to solve the problem of allocating resources to a portfolio of projects, an n-p complete problem (Garey et al. 1976). See Fig. 16. Let us assume an organization that wants to carry out different projects, so that each one will have its objectives, expected profitability, priorities, needs, dates, etc. At the same time, the organization will have limited resources, both personal and material, to try to undertake part of the projects that are available in the portfolio. The available resources are also individual, each with its skills and efficiency in each one of the projects.

The market metaphor in this case is to assume that each of the projects will be represented by an artificial computer agent who will be provided with limited funds that it will use to maximize its utility (profit), to complete the project in time and with the least possible cost.

Each project demands the necessary resources during certain time slots (slots). On the other hand, the resources will also be represented by artificial agents that try

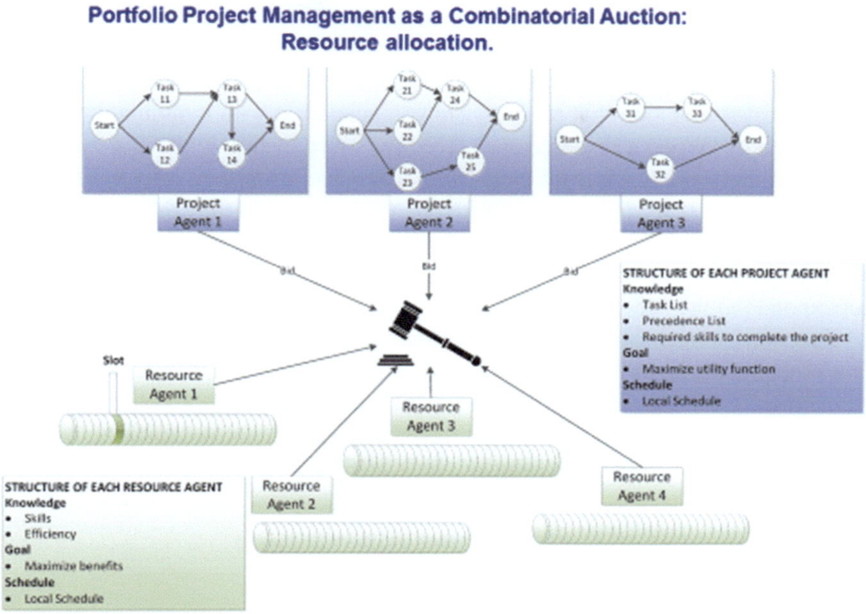

Fig. 16 Management of a portfolio of projects as a combinatorial auction

to maximize their profit; that is, to extract the maximum amount of wealth to the projects and for this they will be willing to sell their temporary slots (ask) to that project willing to pay more (bid). We can formulate the coordination's system as an auction. The resources auctioned are not homogeneous and they frequently are complementary. The value for a buyer of a resource at a given time depends also of having another related resource. The CDA auction will not be valid for this problem. The usual approach is to allow agents to simultaneously bid for various assets. This leads to a Combinatorial Auction (CA). Among the different possible random auctions, we will use an iterative auction to fix prices.

The process is as follows. For a detailed explanation, see Arauzo et al. (2009) and (2010). Each activity is associated with a type of skill and each resource has a set of skills and efficiency in each of them. The greater resource efficiency, the shorter duration is required to complete each task. Projects have precedence relationships from end to end such that a task cannot be started until the precedents have been completed. Resources have their own cost rate. There is one agent-project for each project in the portfolio. Each agent-project, will request the set of resources slots that allow them to achieve their goals at minimum cost. The total cost of the project will be the sum of the prices of the slots of the resources plus an additional penalty in case the project is delivered with an allowable delay. In order to make the bid, the project-agents use a dynamic programming algorithm that evaluates the possible combinations of slots that allow the achievement of the project (Wang et al. 1997).

Since the proposal of activities of project-agent is decentralized and each one seeks its own objectives, the result of all proposals frequently results in incompatible programs that request some resources at the same time and are globally not optimal. The rules of the auction that reduce inconsistencies start with a minimum price for each time slot. When a resource-agent receives more than one offer for one of the slots, it raises its price, while the slots without demand lower their price until a stable price is reached. Once the prices adjusted by the resource-agents, the project-agents renew their local programs according to the new price information, to maximize again their individual profits. The auction continues indefinitely. This procedure corresponds to the sub-gradient optimization algorithm (Zhao et al. 1999).

This non-hierarchical MAS approach has certain advantages in solving allocation problems. It is very flexible and robust to changes in the number of agents of both types, the communication between the agents is minimal and generally, it allows finding solutions very satisfactory from the iterative solution of local problems, which is very interesting in many applied problems.

6 Conclusions

A historical perspective of the Economy in the last 60 years allows us to select some pending challenges and advances in Economics. When we consider Economics as a science that tries to understand how growth generates and it is distributed, we face the first challenge: There has been steady growth but increasing inequality as well. One of the causes of inequality was already point out in by Solow's (1957) seminal paper. There is a welcome residual due to the value added produced by learning, externalities and technological, managerial and institutional improvements. However, since the nineteens, corporate profits are growing over labor income, increasing inequality. The inequality gap has become even greater due to the 2008 crisis. The financial and real sector (FIRE) pumps wealth from the productive sector. To correct inequality, that hinders the rate of growth and undermines democracy, we need measures far beyond Economic Policy. We need deep reforms in Political Economy that let the workers share in a fair way the residual, firm governance and risk. This means deep reforms in the labor market and the FIRE sector.

From the seventies crisis we learned some fundamental messages. (i) Economic Theory is contextual and Economists can learn from different context. (ii) They do not have to wait for a crisis to change their models. Friedman anticipated the explanation of the crisis. (iii) Economic models should specify citizen's expectations. (iv) Adaptative "fast and frugal" expectations, could do as well as the so-called rational expectations, as anticipated almost 35 years ago by Hernández-Iglesias and Hernández-Iglesias (1981). (v) Higher oil prices triggered the crisis but the persistent causes were wrong monetary expansion policies. Just as it happened in the Japanese crisis in 1990, wrong monetary and fiscal policies reinforced the crisis.

The actual crisis lessons. In 1985, something new happened in the history of the Economy. The time series line of the credit that goes to investments in the FIRE sector crosses the line of the credit that goes to investments in the productive sector. The ratio of credit to the FIRE versus credit that goes to the productive sector was 1. In 2015 the ratio was 5. (i) This means a tremendous credit inefficiency. (ii) The bulk of the FIRE sector credit is created by the shadow banking. When citizens accept securitized credit, they become "bankers" without being aware of it. (iii) The FIRE sector is pumping wealth, through mortgages mainly, from the productive sector and families without rents. (iv) This huge transfer of wealth to capital rents has widened the inequality gap, thinning the middle class, and increasing the employment duality. (v) Econometric models, including the DSGE model did not consider the FIRE sector. (vi) The crisis has favored new independent institutions, as INET, to promote rethinking Economics. Thanks to INET, accounting models of the FIRE sector, buried by the orthodox economists, can reveal the cause and persistency of the crisis, and Agent Based (Artificial Economics) Macroeconomic Models will be soon available. (vii) The crisis makes inescapable a deep reform of the FIRE sector that goes far beyond prudential rules and monetary and fiscal policies.

In the last part of the chapter, we discussed what is wrong with Economics as a social science and the promise of the new avenues of EE and AE. Here are some of the conclusions. (i) EE that was developing since the sixties is nowadays accepted as a toolkit to improve Economics in all their subfields. (ii) It made possible to give up some of the unrealistic assumptions of Mathematical Economics and to enrich Economics with Psychology and Social thinking. (iii) EE brought up the need to consider the "value window" (constructivist rationality) and the "exchange window" (ecological rationality) in Economics confirming by experiments the Hume-Hayek hypothesis of the markets as an engine of knowledge and collective intelligence. (iv) Collective intelligence explains that fast and frugal rationality can be as good as full rationality, i.e., that adaptative expectations can match rational expectations. (v) EE provides an internal consistency test for standing economic relationships or models, which may have past the external econometric testing.

AE extends all the findings of EE since by defining soft agents it is possible to simulate the ultimate details of the model dynamics, in terms of the agent's endorsements, the institution and the environment. AE can be a useful tool to *generate* models in many fields of Management Engineering and in any physical landscape populated by social agents. Socially inspired methods such as auctions models can provide solutions to mathematical complex problems (n-p hard) such as scheduling and slot allocations for airlines or to complex problems in Organization such as the supply chain management or the efficient management of a portfolio of projects.

Acknowledgements Gratitude is due to those with whom I shared new ideas in Social Simulation: Bruce Edmonds, Rosaria Conte (†), Nigel Gilbert, Scott Moss and Klaus Troitzsch and in Artificial Economics, Phillipe Mathieu and Frederic Amblard. I also owe gratitude to those at InSiSoc, who took the risk, investing in the challenges and opportunities of ABM applied to Social

Simulation and Economics: Ricardo del Olmo Martínez, Adolfo López-Paredes, José. Pérez-Ríos, Javier Pajares, Segismundo Izquierdo, José M. Galán, Luis Izquierdo, Marta Posada, Alberto Arauzo, David Poza and Cristina Ruiz.

The last part of the chapter draws from research financed by the Spanish Ministry of Science and Innovation, Ref. CSD2010-00034 (SimulPast Consolider-Ingenio 2010) and from EURO-CONTROL-SESAR European Union.

References

Abramovitz, M. (1956). Resource and output trends in the United States since 1870. In Resource and output trends in the United States since 1870. NBER (pp. 1–23).

Albensi, S., Gregory, V., Patterson, C. & Sahin, A. (2013). *Is job polarization holding back the labor market?* New York Federal Reserve Bank, Liberty Street Economics.

Arauzo, J. M., Galán, M., Pajares, J., & López-Paredes, A. (2009). Multi-agent technology for scheduling and control projects in multi-project environments. An auction based approach, *Inteligencia Artificial, 13,* 12–20.

Arauzo, J. A., Pajares, & A. López-Paredes, A. (2010) Simulating the dynamic scheduling of project portfolios. *Simulation Modelling Practice and Theory, 18,* 1428–1441.

Arrow, K. J., Forsythe, R., Gorham, M., Hahn, R., Hanson, R., Ledyard, J. O., et al. (2008). The promise of prediction markets. *Science-New York Then Washington-, 320*(5878), 877.

Balke, T., & Gilbert, N. (2014). How do agents make decisions? A survey. *Journal of Artificial Societies and Social Simulation, 17*(4), 13.

Bernanke, B. S. (2013, November). The crisis as a classic financial panic. Speech delivered at the Fourteenth Jacques Polak Annual Research Conference sponsored by the International Monetary Fund, Washington, November (Vol. 8).

Bertsekas, D. P. (1990). The auction algorithm for assignment and other network flow problems: A tutorial. *Interfaces, 20,* 133–149.

Bezemer, D (2012a). Finance and growth: When credit helps, and when it hinders. In INET Conference "Rethinking Economics and Politics" in Berlin. URL http://ineteconomics.org/conference/berlin/finance-and-growth-when-credit-helps-and-when-it-hinders (last visited, September 2016).

Bezemer, D. (2012b). The economy as a complex system: the balance sheet dimension. *Advances in Complex Systems, 15*(supp02), 1250047.

Bezemer, D. (2014). Schumpeter might be right again: the functional differentiation of credit. *Journal of Evolutionary Economics, 24*(5), 935–950.

Bezemer, D. (2016). Towards an 'accounting view' on money, banking and the Macroeconomy: history, empirics, theory. *Cambridge Journal of Economics, 40*(5), 1275–1295.

Bezemer, D., Grydaki, M., & Zhang, L. (2016). More mortgages, lower growth? *Economic Inquiry, 54*(1), 652–674.

Bezemer, D., & Hudson, M. (2016). Finance is not the economy: Reviving the conceptual distinction. *Journal of Economic Issues, 50*(3), 745–768.

Bezemer, D., & Samarina, A. (2016). Debt Shift, Financial Development and Income Inequality in Europe. Research Report 16020-GEM, University of Gröningen.

Blanchard, O. (2009). *The crisis: Basic mechanisms and appropriate policies*. IMF WP/09/80.

Blanchard, O., & Johnson, D. (2013). Macroeconomics. Pearson (6th ed., p. 246).

Caballero, R. J., & Lyons, R. K. (1990). Internal versus external economies in European industry. *European Economic Review, 34*(4), 805–826.

Cagan, P. (1956). The monetary dynamics of hyperinflation. In M. Friedman (ed) *Studies in the quantity theory of money*.

Chakravarty, S., Friedman, D., Gupta, G., Hatekar, N., Mitra, S., & Sunder, S. (2011). Experimental economics: A survey. *Economic & Political Weekly, 46*(35), 39.

Chen, S. H. (2000) Toward an agent-based computational modeling of bargaining strategies in double auction markets with genetic programming. In *International conference on intelligent data engineering and automated learning* (pp. 517–531). Berlin, Heidelberg: Springer.

Chamberlin, E. H. (1948). An experimental imperfect market. *The Journal of Political Economy,* 95–108.

Clemente, F., Blair, H., & Nick Trokel, N. (2016). Corporate tax chartbook. EPI September 19. http://www.epi.org/publication/corporate-tax-chartbook-how-corporations-rig-the-rules-to-dodge-the-taxes-they-owe/. Last visited September 2016.

Conte, R. (2002). Agent-based modeling for understanding social intelligence. *Proceedings of the National Academy of Sciences, 99*(suppl 3), 7189–7190.

Davis, D. (2016). Experimental methods for the general economist: Five lessons from the lab. *Southern Economic Journal, 82*(4), 1046–1058.

Dilaver, Ö. & Gilbert, N. (2016). Agent-based Macreoconomic Models: An anatomical review. Project.

Friedman, M. (1968). The role of monetary policy. *The American Economic Review, 58*(1).

Galán, J. M., López-Paredes, A., & Del Olmo, R. (2009). An agent-based model for domestic water management in Valladolid metropolitan area. *Water Resources Research, 45*(5).

Garey, M. R., Johnson, D. S., & Sethi, R. (1976). The complexity of flowshop and jobshop scheduling. *Mathematics of operations research, 1*(2), 117–129.

Ghahramani, Z. (1998). Learning dynamic Bayesian networks. In *Adaptive processing of sequences and data structures* (pp. 168–197). Berlin, Heidelberg: Springer.

Gode, D. K., & Sunder, S. (1993). Allocative efficiency of markets with zero-intelligence traders: Market as a partial substitute for individual rationality. *Journal of Political Economy,* 119–137.

Granger, C. W. J. (1969). Investigating causal relations by econometric models and cross-spectral. *Econometrica 37,* 424–438.

Hayek, F. A. (1945). The use of knowledge in society. *The American Economic Review, 35*(4), 519–530.

Hernández, C., Zarrop, M., & Becker, R. (1979). Short term forecasting with the survey of business intentions of the Spanish ministry of industry. In *Jenssen, J., Pau, L. & Straszak Models and Decision Making of National Economies* (pp. 263–271). North Holland.

Hernández-Iglesias, C., & Hernández-Iglesias, F. (1981). Causality and the independence phenomenon: The case of the demand for money. *Journal of Econometrics, 15*(2), 247–263.

Hernández, C., & López-Paredes, A. (1999). Beyond experimental economics. Trading institutions and multi-agent systems. In: *Proceedings of the 5th international conference of the society for computational economics*, Boston.

Hernández, C., & López-Paredes, A. (2000). Engineering market design and computational organization. Growing the market from the bottom-up. In: *Proceedings of the workshop on complex behaviour in economics, CEFI,* Aix-en-Provence, France.

Hudson, M., & Bezemer, D. (2012). Incorporating the rentier sectors into a financial model. *World Economic Review, 1*(1), 1–12.

Keen, S. (2009). Is Rudd the new Whitlam. Debtwatch, March 2 http://www.debtdeflation.com/blogs/2009/03/02/debtwatch-no-32-is-rudd-the-new-whitlam/comment-page-3/. Last visit september 2016.

Legarde, C. (2016). Interview. Q&A With Christine Lagarde: Finance's Firefighter Wants to Be Its Architect. Micklethwait, J. Bloomberg. http://www.bloomberg.com/features/2016-christine-lagarde-interview/. Last visit September 2016.

Ljung, L. (1999). *System identification: Theory for the user* (2nd ed.). Englewood Cliffs, NJ: PrenticeHall.

López-Paredes, A. (1999). Análisis e Ingeniería de las Instituciones Económicas. Una metodología Basada en Agentes. Also publised in the book series by Servicio de Publicaciones de la Universidad del País Vasco. Bilbao (Spain) (2001)

López-Paredes, A. (2000). *Análisis e Ingeniería de las Instituciones Económicas. Una Metodología Basada en Agentes*. Ph.D. Thesis. University of the Basque Country.
López-Paredes, A., & del Olmo Martínez, R. (1998). The social dimension of economics and multiagent systems. In B. Edmonds, & K. Dautenhahn, (Eds.), *Socially Situated Intelligence: a workshop held at SAB'98,* August 1998, Zürich. University of Zürich Technical Report (pp. 73–79). http://cfpm.org/cpmrep44.html. Last visit September 2016.
López-Paredes, A., Hernández-Iglesias, C., & Gutiérrez, J. P. (2002). Towards a new experimental socio-economics: Complex behaviour in bargaining. *The Journal of Socio-Economics, 31*(4), 423–429.
López-Paredes, A., Saurí, D., & Galán, J. M. (2005). Urban water management with artificial societies of agents: The FIRMABAR simulator. *Simulation, 81*(3), 189–199.
Lucas Jr, R. E. (1993). Making a miracle. *Econometrica: Journal of the Econometric Society,* 251–272.
Marcet, A., & Sargent, T. J. (1989). Convergence of least squares learning mechanisms in self-referential linear stochastic models. *Journal of Economic Theory, 48*(2), 337–368.
Muth, J. F. (1961). Rational expectations and the theory of price movements. *Econometrica: Journal of the Econometric Society,* 315–335.
National Commission on the Causes of the Financial and Economic Crisis in the United States. (2011). Final Report of the National Commission on the Causes of the Financial and Economic Crisis in the United States.
Nelson, C. R. (1972). The predictive performance of the FRB-MIT-Penn model of the US economy. *American Economic Review, 62,* 902–917.
Parkes, D. C., & L. H. Unga, L. H. (2001). An auction-based method for decentralized train scheduling. *Proceedings of the fifth international conference on autonomous agents* (ACM, New York, NY, pp. 43–50).
Phelps, E. S. (1967). Phillips curves, expectations of inflation and optimal unemployment over time. *Economica,* 254–281.
Phillips, A. W. (1958). The relation between unemployment and the rate of change of money wage rates in the United Kingdom, 1861–19571. *Economica, 25*(100), 283–299.
Pindyck, R. S., & Rubinfeld, D. L. (1995). *Microeconomics*. USA: Prentice-Hall Inc.
Posada, M., Hernández, C., & López-Paredes, A. (2006a). Strategic behaviour in continuous double auction. In *Advances in Artificial Economics* (pp. 31–43). Berlin, Heidelberg: Springer.
Posada, M., Hernandez, C., & López-Paredes, A. (2006b). Learning in continuous double auction market. In *Artificial Economics* (pp. 41–51). Berlin, Heidelberg: Springer.
Posada, M., Hernandez, C., & López-Paredes, A. (2007). Emissions permits auctions: an agent based model analysis. *Social Simulation: Technologies, Advances and New Discoveries: Technologies, Advances and New Discoveries,* 180.
Posada, M., Hernández, C., & López-Paredes, A. (2008). Testing marshallian and walrasian instability with an agent-based model. *Advances in Complex Systems, 11*(02), 249–260.
Posada, M., & López-Paredes, A. (2008). How to choose the bidding strategy in continuous double auctions: imitation versus take-the-best heuristics. *Journal of Artificial Societies and Social Simulation, 11*(1), 6.
Posada, M., & Hernández, C. (2010). The effect of transaction costs on artificial continuous double auction markets. In *Progress in Artificial Economics* (pp. 65–74). Berlin, Heidelberg: Springer.
Rassenti, S. J., & Wilson, B. J. (2004). How applicable is the dominant firm model of price leadership? *Experimental Economics, 7*(3), 271–288.
Rassenti, S. J., Smith, V. L., & Bulfin, R. L. (1982). A combinatorial auction mechanism for airport time slot allocation. *The Bell Journal of Economics,* 402–417.
Romer, P. M. (1990). Endogenous technological change. *Journal of Political Economy, 98*(5 pt 2).
Sachs, J. (2010). America's political class struggle. *The project syndicate*.
Sargent, T. J. (1993). *Bounded rationality in macroeconomics*. Oxford: Oxford University Press.
Schasfoort, J., Godin, A., Bezemer, D., Caiani, A. & Kinsella, S. (2016). Interacting prudential and monetary policies: Financial stability from the bottom up. AE 2016 Conference, Rome.

SESAR Joint Undertaking. (2014). Agent-based pilot model for alternative primary airport slot allocation with price-setting auctions. Fourth SESAR Innovation Days. Ramírez, M., Villafáñez, F., Arauzo, A. & López-Paredes.

Shu, H. C. (2012). Varieties of agents in agent-based computational economics: A historical and an interdisciplinary perspective. *Journal of Economic Dynamics and Control,* 36(1), 1–25, 20.

Simon, H. (1982). *Models of bounded rationality* (3rd ed.). Cambridge: MIT Press.

Skidelsky, R. (2015). *Keynote at the teaching economics in the 21st Century conference*. Berlin School of Economics and Law.

Izquierdo, L., & Izquierdo, S. (2015). *Artificial economics: What, why and how. Based on keynote artificial economics conference, AE 2015*, Porto. http://www.luis.izqui.org/papers/artificial-economics.pdf. Last visit September 2016.

Smith, A. (1776). *An inquiry into the nature and causes of the wealth of nations*. Publisher/Edition. London: Methuen & Co., Ltd. Pub. Date. (1904), 66.

Smith, V. L. (1962). An experimental study of competitive market behavior. *The Journal of Political Economy,* 111–137.

Smith, V. L. (1976). Experimental economics: Induced value theory. *The American Economic Review, 66*(2), 274–279.

Smith, V. L. (1982). Markets as economizers of information: Experimental examination of the "Hayek hypothesis". *Economic Inquiry, 20*(2), 165–179.

Smith, V. L. (1989). Theory, experiment and economics. *The Journal of Economic Perspectives, 3* (1), 151–169.

Smith, V. (2007). *L.* Rationality in economics: Constructivist and ecological forms. Cambridge University Press.

Smith, V. L (2008). Experimental economics. In S. N. Durlauf, & L. E. Blume (Eds.) *The New Palgrave Dictionary of Economics* (2nd ed.).

Solow, R. M. (1957). Technical change and the aggregate production function. *The Review of Economics and Statistics,* 312–320.

Soros, G. (2012) *INET speech at opening session*. Berlin Conference.

Sunder, S. (2004). Markets as artifacts: Aggregate efficiency from zero-intelligence traders. In M. E. Augier & J. G. March (Eds.), *Models of a man: essays in memory of Herbert A. Simon*. Cambridge: MIT Press.

Thompson, D. (2013). The Atlantic. https://www.theatlantic.com/business/archive/2013/12/the-most-important-economic-stories-of-2013-in-44-graphs/282193/. Last visit: November 2016.

Wang, J., Luh, P., Zhao, X., & Wang, J. (1997). An optimization based algorithm for job shop scheduling, *Sadhana, 22*, 2.

Wellman, M. P., Walsh, W. E., Wurman, P. R., & MacKie-Mason, J. K. (2001). Auction protocols for decentralized scheduling. *Games and Economic Behavior, 35*(1), 271–303.

Wittenmark, B. (1974). A self-tunning predictor. *IEEE Transactions on Automatic Control 19*(6).

Zhao, X., Luh, P. B., & Wang, J. (1999). Surrogate gradient algorithm for Lagrangian relaxation. *Journal of Optimization Theory and Applications, 100*(3), 699–712.

Chapter 4
The New Industrial Organization

Ricardo del Olmo Martínez and Adolfo López-Paredes

Abstract Economic Activity takes two alternative forms: the Market and the Firm. Economics is a social science that tries to explain how to generate wealth and how it is distributed. The firm is a social organization whose members decide to cooperate to generate wealth and its distribution among the stakeholders. They certainly share a common goal. However, to translate economic principles to management is an open challenge. Traditional IO as understood among the economists deals with the generation of wealth through the market, which if it is well designed will achieve a fair distribution through endogenous dynamics towards equilibrium. On the other hand, a proper theory of the firm needs explicit rules of governance and operations. This fact requires a New I.O dealing with uncertainty far beyond probability; individual and collective bounded rational agents; specialization and heterogeneity; imperfect information and variety; incentives and penalties to avoid free riding, and how to develop core competences such as entrepreneurship, innovation and knowledge management. The paper revised I.O and ends up with a map of Management Sciences to help designing the Management Engineering curricula and the range of specific skills and competences demanded by different institutions.

Keywords Industrial organization · Economics and management of the firm · Governance and strategy · Organizational sciences

---

R. del Olmo Martínez (✉)
Grupo INSISOC Dpto. de Ingeniería Civil, Escuela Politécnica Superior
Universidad de Burgos, c/Villadiego S/N, 09001 Burgos, Spain
e-mail: rdelolmo@ubu.es

A. López-Paredes
Grupo INSISOC Dpto. de Organización de Empresas y CIM. Escuela de Ingenierías Industriales, Universidad de Valladolid, Paseo del Cauce S/N, 47011 Valladolid, Spain
e-mail: insisoc@gmail.com

C. Hernández (ed.), *Advances in Management Engineering*, Lecture Notes in Management and Industrial Engineering, DOI 10.1007/978-3-319-55889-9_4

1 Introduction

There are two alternative forms; two basic institutional frames under which economic activity may be organized: the market and the firm. Although each develops under different organizational patterns, both the firm and the market, function through the (social) interaction of the different economic agents that participate in them and have common goals. In all economic events, there is a dynamic process, which conditions the results obtained both by the firm and by the market. This process is subject to continuous changes and adaptation, due to the business environment and the social nature of the individual participants.

In the market, the free interplay of offer and demand looks after the distribution of the different resources between individuals, leading to the determination of equilibrium prices. In the firm, on the contrary, this task develops through a governance structure that assures management and planning.

The market provides the simplest form of coordination, as its functioning requires no planning of conscious activity on the part of individuals, but it is rather its own individualistic orientation what dynamizes resource assignation. The characterization of the functioning of the market is therefore justified, as a mechanism regulated by an "invisible hand", in so far as, the system of prices is not consciously created (Lipsey 1963).

With such a system, it is not necessary to foresee and to coordinate all the necessary variations. These occur automatically resulting from separate decisions taken by a large number of individuals, who seek their own benefits, but are obliged to respond, in relation to changes in demand and in prices. The firm, however, requires greater personal interaction between its parts, a governance structure, as well as more complex and lasting relations between them.

The development of organizational patterns is similar to the biological process of natural selection. Competition in the market leads to the predominance of more efficient solutions. Unlike the biological case, this social evolution is much more rapid, is transmitted by learning (Hayek 1988), and is subject to human manipulation. Moreover, the opposing organizational patterns can coexist, above all in the short term.

The study of factors that determine the adoption of one or another alternative constitutes the first big question to analyze, as these factors will set the limits and the *raison d'être* of the firm as against the market. Why, on some occasions, there are organizations that mediate in the transactions between individual agents and the market?: the discovery of the firm as an alternative mechanism to the market for resource assignation.

This paper starts with the definition of some exemplary concepts in Economics that constitute the grounding of the different theories on the firm and the market, as alternative institutions that compete between each other. We make a succinct review of the theories that in our view can be considered as the most representative. Among others, some are the result of having studied these core facts of the economic activity with greater or lesser emphasis and degrees of success.

Neo-classicism is locked in the omnipotent authority of the market, the only economic scenario in which the firm is reduced to a mere intellectual concept, a production unit. The institutionalists countered that approach, principally through the Nobel prize-winner, R. Coase, and the undeniably pragmatic observation of market inefficiency, as well as the discovery of the firm as an alternative mechanism to the market for resource assignation.

Management theories, in addition to profit margins, introduce other components like restrictions on the objective function; but it is the behavioural theories that substitute the objective of profit maximization by that of satisfaction under limited rationality within the framework of structures of ownership rights in the firm. Thus, we come to Agency theory with the conflicts of control between ownership and management and the agency-cost minimization objective, which are nothing but a very concrete definition of transaction costs, as opposed to market costs. Finally, we deal with the resource and capabilities view of the firm.

The paper concludes by presenting a structure of integration of the Economy, the Economic Theory of Contracts and the Theory of Endogenous Growth, which Professor C. Hernández refers to as the New Industrial Organization (Hernández 1997).

2 Core Concepts

Organizations are the result of the contractual interaction of human beings who follow a particular conduct. In essence, it is supposed that individuals seek their own self-interest (profit) in an intelligent way and interact in an environment characterized by the cost of information, which is neither perfect and nor asymmetrically held, and by the need to relate to other people. They do it with cognitive limitations, individual bounded rationality.

Besides, although opportunism is a natural consequence of the search for self-interest in an environment with information shortages, opportunism is constrained by its own rationality. This brings up substantive uncertainty that ads to the one coming from organization environment.

In what follows, we shall define a series of core concepts in Economics on which to construct the different theories that exist on the firm and the market, such as competitive alternative institutions, among others.

2.1 Uncertainty

Nothing is more certain than the predominance of uncertainty over the consequences of any economic decision. There is no doubt that uncertainty is inseparable from the human condition and dominates the majority of their thoughts.

Knight (1921) highlighted the failings of the probabilistic approach in characterizing the essential aspects of the business managers, or the entrepreneur's role or the forms evolution: the consideration of uncertainty. Uncertainty per se really introduces something fundamental into Economics: individuals possess different information and their attitudes may change drastically.

Risk (chance) refers to recurrent situations in which, through repeated observations, it is possible to assign frequencies and, assuming an underlying regularity, the corresponding probabilities, for some possible perfectly defined and identifiable results. Uncertainty refers to situations that present no regularity under observation, the results of which are on occasions not clearly identified, and in any case, may not be assessed as probabilities.

In this way, Knight maintained that it is not measurable risk, but uncertainty that is not assessable in terms of probabilities.

However, throughout the Microeconomics literature it is difficult to stumble across this distinction, as the majority of authors consider uncertainty measurable through subjective probabilities. In this way, and with the help of Bayesian statistics, it converts uncertainty into risk, and the treatment of the different situations is simplified (Pindyck and Rubenfield 1998); mindful with the exceptions that Baumol (1961) and Frank (1991) pointed out.

2.2 *Limited Rationality*

The hypothesis that the agents are rational is the central assumption of many theories of the social sciences. Its role is particularly obvious in economic analysis (Kahneman 1994). The term limited rationality is used to designate the rational choice that takes into account the cognitive limitations and calculative capability of the decision-maker, fundamental to estimating the market behaviour of the economy.

The notion of bounded rationality was introduced by Simon (1947), diverting from the theory of the Subjective Expected Utility (SEU) of global maximization, lying beneath neoclassical Economics. It is a consequence of there being empirical knowledge on human thought and the decision process.

Limited rationality is procedural, not substantive. Substantive rationality comes under the neoclassical model, according to which, it is enough to assure compliance between what the behavioural model predicts and what it is really observed. However, human behaviour cannot predict the optimal behaviour in a given environment. It depends on how the economic actors perceive and represent the environment, how they define their goals and the methods to value the achievement of those goals, what facts they know or assume and what strategies they have to resolve the problems.

The rationality of the economic actors, can be defined by the process that they use to construct their decisions, but that process cannot be assumed from the description of the objective that is pursued when the problem is solved. It cannot be determined inductively from empirical observation, or inferred from behavioural theories with an empirical base. In particular, as Simon (1997) stated, the economic agents will be highly influenced by social change.

Thus, instead of searching for the optimal, the cost-effective and the difficult solution in accordance with their calculative capability, individuals are content to find satisfaction by renouncing the optimal. The agents, aware of their limited rationality, act by trying to do it as best as they can, given the limitations under which they are working. And they also learn and use frugal and fast rules (Posada and López-Paredes 2008; Gigerenzer et al. 2002; López-Paredes et al. 2002).

In spite of it all, individuals are complex entities that will not always follow rational criteria when taking their decisions. Apart from the informational limits, the deviations of the supposed rational conduct are due to the information processing it involves. According to the model of rational selection, agents evaluate the events or the sets of events from the point of view of their global influence on their utility function.

However, Kahneman and Tversky (1979) observed that the agents usually weigh up each of the events separately. They developed the prospective theory, to explain these distractions, to eliminate the paradoxes of consistency in the expected utility from consumer behaviour, through a value function defined over changes that the economic agents experience in their wealth.

2.3 *Information Asymmetry*

An efficient choice requires information on individual tastes, technological opportunities and resource availability. All the information is not held by all individuals, nor is it the same for all those who hold it.

Regrettably, information asymmetries are omnipresent in economic relations: "The clients know more about their tastes and inclinations than the firms. Firms know more about their costs than the government, and all agents assume actions that are partially unobservable" Salanié (1997).

The competitive system can be represented, in the same way as any resource assignment mechanism, as an information exchange structure between individual agents. From this point of view, the key to its efficient operation is the transmission of sufficient and identical information for all agents. The fact that the agents hold different information before or after reaching an agreement generates, respectively, events termed in the literature as adverse selection and moral risk, giving rise to opportunistic behaviour that provokes inefficient resource assignation.

Williamson (1985) describes opportunism as a wily search for enlightened self-interest; in other words, individuals can leave contractual commitments unfulfilled, if so required. In this way, when a conflict arises between what people wish to do and what they have accepted to do for others, will act in their own self-interest if it is costly for the other parties to supervise and control their behaviour. In no way does this idea imply that all individuals behave in an opportunistic way, nor that they do so at all times; being difficult to distinguish honest individuals from those who are not and, therefore, to determine when the opportunistic behaviour will take place.

2.4 Specialization, Negotiation and Cooperation

Economic activity and the value it generates comes from productive specialization and the subsequent exchange. Efficient external markets and an internal market organization are required to materialize those potential benefits; through the exchange between links in the value chain, between the divisions of the decentralized corporate firm, or between the components of the virtual firm, as previously pointed out by Hirshleifer (1956) with a keen sense of anticipation.

More and better can be produced when cooperating; each one specializing in their productive activities and then negotiating between each other to acquire those goods and services that are necessary. This specialization in production and exchange should be coordinated by providing cooperation, from motivation and incentives for the participant agents Milgrom and Roberts (1992). But, effective coordination is achieved when they all have access to the necessary information for efficient resource assignation. Thus, there are different organizational structures to achieve that coordination.

The cost of production is reduced and the product value increases by specialization, but generates coordination costs. Among these costs, the most problematic ones are those called motivation costs or incentives, related with the opportunistic behaviour of the actors in the exchange. In more complex products, scale economies and learning emerge. Adam Smith had already highlighted the specialization of functions as a source of efficiency (Friedman 1991).

The modern example of specialization is the value chain and the organization of the firm by processes. The productive process is broken down into stages of activity and the relation of exchange, among which, results into as a bilateral monopoly. From this model, the concept of incentives emerges. Two decades ago, the model of bilateral monopoly was a curious field in our microeconomic culture; however, it has finally assumed greater importance because it applies to many problems of real negotiation and transfer pricing.

The transaction cost economy of Coase, and the economy of transfer pricing of Hirshleifer are two alternative discourses on the difficulties of establishing incentives that reduce the internal exchange costs of the firm. The concept of exchange becomes an internal mechanism of the firm that works with internal transfer pricing. The need therefore arises to include imperfections and asymmetries in that exchange process to set the price for internal transactions in the modern multi-divisional firm: the theory of transaction costs.

However, this is not enough as the following fact proves. Goods exchange generates wealth but as such is mute about its distribution. In fact, consider a firm with two divisions, A and B. If say division A increases efficiency, the optimal transfer price between the two divisions leads to the maximum profit for the firm. However, division A may well have less profit than before, whereas division B will take all the increase in profit. Therefore, incentives and cooperation have to be induced in the multidivisional firm; something else that pricing, cost or information. It is a question of incentives and fairness.

2.5 *Variety*

In the 1930s, General Motors literally pushed Ford out of the market, all because of an idea that it put into practice: "a model for each pocket and for each use: all Americans can have an automobile to match their taste, needs and purchasing power". Specialization and the cost objective are not sufficient to compete. It is a question of generating value from variety: price margin over cost. Organizational changes also become evident: the multidivisional firm, where each division is treated as a different business unit.

This new organizational structure opens the door to external economies based on technological improvements, which opens up another means of wealth generation, somewhat more important to explain the Solow's residual in the long term. So Romer (1990), used the idea of Marshall, to ground the new "theory of growth", according to which knowledge and more specialized machinery can originate externalities. Knowledge may be included as a non-rival factor.

Finally, there are also external economies arising from the variety: external but internal to the multiple divisional firm. For example, the increase in the number of intermediate and increasingly specialized factors, as happens with the multidivisional firm and flexible production systems. Caballero and Lyons (1990) gave further consideration to the importance of market size and technological innovation and diffusion, as elements that generate positive external economies.

2.6 *Exchange: Institutional Dimension of Production*

It no longer makes any sense to talk of product exchanges, except perhaps for raw materials, but instead of the inherent attributes of the product. Productive processes may be broken down into n different activities that may be done at places and under alternative conditions, through independent agents who freely exchange, coordinating without coercion and generating advantages. This coordination of the exchange process entails costs. Firms and markets are two scenarios in which to estimate where the exchange costs are lower.

Exchange entails transaction costs that arise, on the one hand, because of the coordination of offer and demand, in such way that they rationally maximize their personal utilities reassigning the use of goods and resources; and, on the other, the motivations of individuals, due to their possible opportunistic behaviour.

"We can see the firm as a series of markets, throughout its value chain, with their corresponding transaction costs; or the market as a production unit where the factors are the rights on offer and the production of that exchange activity, the satisfied demand, the rights conceded" (Hernández 1997).

Property rights mark the position of each agent in relation to the scarce resources. It is the ownership rights that constitute the importance of the contribution from Coase in 1960, the relevance of social cost (Coase 1960). Exchanges are, in reality, the transference of property rights.

A system of ownership rights configures a certain system of incentives; therefore, substantial variations in ownership rights can change the dynamics of the production system, as well as the hierarchical relations of the firm and, in general, the relations of power or dominance that prevail in society.

Uncertainty, limited rationality and information asymmetries are circumstances associated with the governance mechanism or mode of exchanges in the firm and in the market. These factors imply a certain sort of incomplete contract in all economic relations. In short, there are always transaction costs that limit the initial advantages of pure exchange.

2.7 The "E" Factor and Other Intangibles

Economic growth is not completely explained by the growth of productive factors. In the long run, Solow (1956) found that the increase in wealth could not be attributes to the usual factors of production. There was a welcome unexplained residual. Marginalism could explain this residual. A dummy factor was introduced called Total Productivity Factor. Perhaps should have been a "measure of the economist ignorance". It was important to determine the set of non-physical factors, to understand the nature of the Solow's residual, such as the accumulation of capital, educational improvements, technological change, etc., and finally, if applicable, to guide the growth policy accordingly.

The contributions of the theories of endogenous growth underline that the residual is fundamentally due to the capability to improve the intangibles of the firm, and in particular, knowledge accumulation (Lucas 1993). The conclusion is that management capability to mobilize tangible and intangible resources would be, in the final analysis, the complete explanation of the residual.

The success of the firm depends of the set of its resources, among which of special importance are the training of human resources, business organization and that visible hand (Chandler 1977) that we call entrepreneurship. It is the sixth productive factor, a generator of added value, after land, work, capital, human capital and technology. As Gordon (1993) affirmed, it should now form part of the principles of Economic Theory.

Economic activity may be defined as the transformation of attributes of goods and services through the integration of technology and organization, the task of management; and the change of those physical attributes into ownership rights, into value, which is the task of the entrepreneur.

It is not so important for the entrepreneur to coincide with other functions. It is the entrepreneur who adds value and the manager who reduces costs (efficiency). There are therefore two groups of basic economic agents with different roles—entrepreneurs and managers—although the same person often fulfils both functions.

Managerial functions are: to coordinate production factors to achieve efficiency; to formalize generally incomplete contracts with the members of the firm, determining the incentives; and to capture opportunities for continuous improvement of the business, from a portfolio designed by the entrepreneur.

The role of the entrepreneur can be summarized as: seizing opportunities for gain, exploiting market imperfections; planning and deciding on the business portfolio and its dynamics; and innovating in a changing environment where variety and short product life-cycles prevail.

We can see some of the different approaches to the entrepreneur that we consider relevant. Cantillon (1775) accurately identified the profile of an entrepreneur, by highlighting the nature of the risk that entrepreneurial decisions involve, the innovative function that focuses on market opportunities and the motivations of the entrepreneur. Schumpeter (1954) highlighted the innovative nature of the business activity. Knight (1921) considered that the entrepreneur is a guarantor of rents, while the benefits of the firm represent payment for taking those risks. Kirzner (1997) took imitation and innovation into account. "The role of the entrepreneur is to notice what others may have overlooked" (Kirzner 1982). Casson (1982, 1990) agreed with Kirzner in his appraisal of the entrepreneurial role as gainful re-assignation of resources and of the entrepreneur as an agent of change. He also described the personality and the motivations of the entrepreneur with an economic and psychological analysis.

In spite of what has been said, the role of the entrepreneur is frequently linked to the managers; particularly in the modern corporation, where he appears under different denominations. In the firm that is decentralized into business units, and cooperative entrepreneurial networks (virtual firm), the entrepreneurial activities are localized both within the external market and in the internal market of the firm. The role of entrepreneurs is, even so, carefully define by their participation in the definition of corporate strategy, the portfolio of future business and its proactive approximation to that strategy.

3 Economic Theories of the Firm

The different theories and research programmes that shape the Economic Theory of the Firm may be considered as attempts, at all times insufficient, to accommodate some, or all, of the above-mentioned core ideas the conform the complexity of the Economy.

The analysis of the firm, from the economic point of view, is complicated by dealing with real organizations, and therefore with individuals. The firm is constituted by a set of people with different knowledge and interests, where a response has to be given to the ensuing problems of motivation and coordination, with the aim of achieving the coherent and worthwhile behaviour of all participants (Milgrom and Roberts 1992).

The flow of Economic Analysis has been enriched with contributions arising from the at-all-times unsatisfied need to endow that reality of economic activity with normative and/or positive bases: the agents have limited rationality, subjected, moreover, to institutional restrictions. From that point of view, it is possible to value the achievements and expectations of the different approaches to the Economic Theory of the Firm and the New Industrial Organization in a better way.

Economic analysis provides in this way the fundamental theoretical field to understand more fully the problems of business decisions, both for internal resource assignation as well as in relation to the environment. We can distinguish three economic approaches to the phenomenon of the firm, each one of which offers interesting applications. These are conventional Neoclassical Analysis, Contractual Theory and the Evolutionary Theory of the Firm.

Conventional Neoclassic Analysis sees the organizational dimension of business reality completely in the abstract, which has a twofold consequence. On the one hand, it allows us to undertake an analysis of the markets and to construct a theory of price formation, capable of explaining the logic of business survival (Alchian 1950). On the other hand, it sterilizes theory for its use in internal decision taking.

The second approach-Contractual Theory-emerges to endow Neoclassical Analysis and its application with greater realism, the origin of which may be found in the work of Coase of 1937. It explains the existence of the business form of economic organization based on the transaction costs inherent to the exchanges completed on the market. On the one hand, the Theory of Ownership Rights analyzes how the nature of contracts conditions the behaviour of economic subjects. On the other, Agency Theory provides an analytical framework that gives insight into the contractual relations between the participants of the firm, in their positive aspect, and seeks to serve as the foundation for the development of contractual schemes for the agent-principal relation, such that they are optimal for that contractual development, under prevailing norms.

With a process approach, the third approach defends the firm as a unit of accumulation of knowledge and capabilities. Knudsen (1995) sustains that although it coincides with neoclassical theory, by considering the firm as a unit of production, it has in reality implied a gradual enlargement of the assumptions of market behaviour.

The most significant features of the neoclassical model, of transaction cost theory, of ownership rights theory, agency theory and evolutionary theory are presented in the following section.

3.1 Neoclassical Theory

Over some time, the thoughts of economists have centred mainly on the problems of scarcity and wealth; in other words, those relations that in a unilateral way link people with things. Value, exchange, production and the market were the fundamental concepts on which their interest centred.

The firm, excluded from this theoretical framework, was assimilated with production, understood as the change of factors in products, with the help of a transformation process belonging to each industry and in accordance with the state of the technology. The sole agent of multilateral cooperation of interest to the orthodox economist of the 18th and 19th century was the market, and the essential economic decision referred to commercial exchange.

Subsequent developments, completed towards the end of the 19th century and in the first decades of the 20th century, although clarifying some aspects and enlarging the perspective of classic microeconomy, maintained, in essence, the same earlier model of the firm.

Although no authentic Economy of the Firm existed, a microeconomic theory of the firm had formed in the neoclassical framework: that which studied the behaviour of productive economic units that operated under very restrictive suppositions, which together are all known as the market economy (Naylor and Vernon 1969).

We may say that the neoclassical model is a description of the market, the framework in which the firm operates, more than the firm itself. Its operation and the relations that are formed within it are ignored in this way, as well as the reasons that explain its existence. In this environment, the invisible hand of Adam Smith is at its most meaningful, in so far as the prices transmit the necessary information to all the agents, so that efficient assignation of all resources will ensue from the optimization of their particular wealth.

In this respect, assuming that profit maximization is the sole objective of the firm is a deduction based on the action and the operation of the market, and not of a specific study of the real motivations of the business. The methodological defence of such an assumption is based on greater interest because of predictive power than because of the realism of the hypotheses. We may therefore suppose that the entrepreneur acts as if seeking to maximize returns, as otherwise, competition would have to shift to the market.

Economic analysis has frequently been criticized because of its shortcomings with regard to the management of the firm. Thus, for example, in the productive area, the Theory of the Firm considers that the achievement of productive efficiency is a technical reality: the production function is an optimum technology. However, it is a fact that most business management problems consist of the search for efficient resource assignation, which is of course available through that technological optimum.

That critical comment is nonetheless considerably Byzantine, as it accuses Economic Analysis of missing objectives that are not its own. As Jensen (1983) thoughtfully points out, among many others "… the literature that falls under the heading of 'Theory of the Firm' is not a positive theory of the firm, but a market theory".

Despite these criticisms, the neoclassical model offers at least four sets of fundamental contributions for an economic theory of the firm: (1) it supplies the conceptual framework of internal resource assignation process; (2) the microeconomic analysis offers the scientific basis for decision-making on the relations of the firm with the environment; (3) it provides the fundamental concepts used in the formalization of multiple decision problems; and, (4) it provides an approach to the study of human behaviour.

3.2 Transaction Cost Theory

Since the 1930s of the last century, but fundamentally during the seventies, this panorama changed radically. Holistic methodology present in the earlier stage gave way to individualistic concepts, but in accordance with the social nature of the economic discipline. The development of management and, above all, behavioural theories prompted the institutionalists, following the pioneering work of Coase (1937), to develop concepts of the firm as an alternative to the market, and that provoked important reviews of the ideas on how and why firms operate.

The earlier attempts to extend the market to the point of it subsuming the role of the firm are radically improved by the contribution from Coase, in disagreement with the explanations on the existence of the firm, which always overlooked their nature. He was not in agreement with the view of the firm as a black box into which streams of productive factors entered and a flow of products came out, as this conceptualization made an abstraction of very important components of the real world, which missed the essence of the nature of the firm.

In the context of the theory of general equilibrium, it is very difficult to justify the true existence of firms, as all interactions are done through the price system. As Coase (1937) highlighted: "The hallmark of the firm is the suppression of the price mechanism". Information asymmetry is a further challenge to take into account in the general model.

Firms emerge provided that the benefits of coordinated team production exceed those derived from the formalization of individual contracts. The costs of information, and contract negotiation and implementation are not insignificant. Costs that Coase termed transaction costs. Given that the firm contracts in exchange for a salary that is set in advance, the need for supervision and control emerges. However, the costs of supervision and monitoring are different for each firm and quickly grow with size. Therefore, the volume of these costs is an indicator of the size of the firm.

There are therefore some costs that arise from the use of the price system, which are incurred when conducting transactions in the market; when making use of the price mechanism. In consequence, the firm is an alternative mechanism to the market and arises because it is able to organize certain activities in a more efficient way than through the market.

The firm, for Tirole (1989), appears because it is capable of producing or selling more efficiently than its component parts could do separately. There are two reasons that justify the above. The first is that the firm uses synergies between different units to exploit economies of scale and reach, faced with the indivisibility of certain factors in the production of one or more products. The second reason is that transaction costs have their origin in market imperfections. The firm never supplants the market as a mechanism for assignation, but replaces it in those activities in which the firm can achieve more efficient assignations: minimizing production and transaction costs.

There are some antecedents of the transaction cost economy in the economic (Knight 1921; Commons 1934; Coase 1937), the legal (Llewellyn 1931) and the organizational (Barnard 1938; Simon 1947) field. However, Williamson (1975, 1985) rediscovered and developed the transaction cost model. The economy of transaction costs adopts a contractual approach in the study of economic organization. It maintains that any question that may be expressed as a problem of contracting can be studied in terms of the economy of transaction costs.

In comparison with other theoretical proposals that exist for the study of the organization, Williamson (1986) considered that the economy of transaction costs is more microanalytical. It is more realist in its hypothesis of behaviour; it introduces and develops the economic importance of the specificity of assets; it resides more in comparative institutional analysis; it makes reference to the firm as a governance structure more than as a production function; and it gives greater importance to the ex-post institutions of the contract, with special emphasis on private order.

The relations between economic agents may be better explained by considering the transaction as a unit of analysis; a concept that includes both exchanges and contracts. Exchange would be total transference of ownership rights on a resource that implies no future promises or responsibilities. A contract is the promise of a future result, because one of the parties makes an investment the profitability of which depends on the future behaviour of the other party.

Transaction costs have their origin in the establishment of the conditions of exchange, and are of two types: the ex-ante costs and the costs of the actions and tasks that take place when establishing the contract (negotiation, drafting and guarantee of the agreement), and the ex-post costs or those due to administration, obtaining information, supervising and obliging compliance with the conditions of the contract.

Likewise, Williamson (1985) points to two groups of fundamental conditions that have to arise simultaneously for there to be transaction costs. Conditions related to the behaviour of the individual, especially those associated with the limited rationality of human beings and the opportunism of economic agents. Conditions that are related to the environment of the transaction, in particular, uncertainty over the future and the habitual existence of reduced groups of agents with which to enter into contracts.

In addition, we should take into consideration the specificity of the assets involved in the transaction. In a transaction, it is considered that an asset is specific when it can not be reassigned for an alternative use without a significant reduction in its value. This specificity is considered the most significant dimension in the definition of a transaction, but it is not the only one, as the uncertainty that surrounds the transaction and the frequency with which it is done will also have to be considered.

Transactions are regulated by contracts, the different types of which give rise to different forms of managing transactions in accordance with their characteristics.

Even though firms and markets represent alternative ways of organizing transactions, Putterman and Kroszner (1986) considered that market economies constitute a single network of the entire fabric of the economy. In other words, markets may also exist within organizations and, in turn, the markets may up to a certain point be organized. As Leibenstein (1987) and Douma and Schreuder (1998) pointed out, in practice, markets and organizational coordination are often found in combination.

3.3 Theory of Ownership Rights

The Theory of Ownership Rights (Demsetz 1967), initially formulated by Alchian and Demsetz (1972), provides explanations that justify the firm and the figure of the entrepreneur, but where the relation of authority is not the entrepreneurial. In addition, it seeks to explain not only the existence of the firm, but also its structure.

From that perspective, the firm is contemplated as a set of participants in productive cooperation, who find themselves in a situation that is characterized by the existence of an agent. That agent occupies a central position by participating in the contracts of the other inputs and, hence, that is termed a system of team production. As Alchian and Woodward (1988) affirmed, the figure of the administrator or director of overall production is emphasized.

There is production in a team when the set of individuals cooperate by using different resources to arrive at a product that does not correspond to the sum of the separable outputs for each resource that is employed. Moreover, the set of resources that are used belong to no one single person. Under these conditions, there are problems of measurement and monitoring of individual performance, as well as incentives, as it is costly to determine the contribution of each member of the final product.

On this basis, the contents of the rights over people affect the assignation and use of the resources in a specific and predictable form. Therefore, the effects arising from the possible assignation of ownership rights on economic activity may be determined. These possibilities lead to different structures for payment and sanctions, as Furubotn and Pejovich (1972, 1974) affirmed, determining the conduct of participants. In this way, an interconnection between rights, incentives and behaviour is brought to light.

The market is not an efficient form of governance for team production, as no modification is introduced in the incentives of the participants. Moreover, the lower the remuneration demanded by a possible team member, the greater the incentives to reduce the level of performance.

On the contrary, the firm organizes production in a team efficiently, by establishing an agent, whether an individual or a group, in charge of measuring the productivity of the other members and to remunerate them or sanction them accordingly. That agent is the entrepreneur or the director, who when negotiating prices with the owners of the inputs, as well as directing and observing the actions of employees and the use of the inputs, will do nothing else than measure the productivity of each resource and remunerate it in consequence.

The appearance of the entrepreneur in the team modifies the incentive of its members when rewarding performance. Reviewing and closing contracts, so that the best can be selected and each member rewarded according to their productivity. In addition, the performance of the team depends on the actions of the entrepreneur, who should find incentives for them to perform their role. It is precisely the residual earnings they receive, as well as the possibility of selling their position in the team that constitute the incentives to do so.

In conclusion, according to Alchian and Demsetz (1972), and under a criteria of economic rationality, the firm is born through two conditions that arise simultaneously, when: (1) it is possible to increase the global productivity of a set of resources through the establishment of a system of team production; and, (2) the cost of disciplining the team members through the figure of the entrepreneur does not exceed the earnings in global productivity that are obtained through the formation of the team.

The above arguments have received numerous criticisms, especially those that defend more participative forms of organization. In this sense, Putterman and Kroszner (1986) warned that it is not necessarily the right to the residual earnings of the central agent that leads it to act in more efficient way, as the costs implied by this class of coordination are higher than what is achieved in exchange.

These forms of cooperativism would partly explain the existence of a multitude of programmes, such as the Israeli kibbutz and industrial complexes like Mondragón, which, through greater worker participation in the decisions that affect them, can improve individual performance without resorting to the production team approach of Alchian and Demsetz. In addition, the benefit or residual earnings to which reference is made depend more on circumstances and not only on the level of production.

3.4 Agency Theory

Organizations, in general, and the firm, in particular, are complex and are constituted by a multitude of heterogeneous agents. Therefore, a transaction costs analysis advises defining archetypes that can channel contractual relations and the costs that correspond to the two agents: the principal and whoever is strictly speaking the agent.

Agency Theory arises within the institutionalist economy, for Jensen and Meckling (1976), on a different branch from transaction costs branch, because of the interest in the analysis of contracts between individual economic agents. Their object is to minimize agency costs that arise from all forms of cooperation between two or more people. The agency relation appears when a person, the principal, commissions another, the agent, in exchange for a remuneration, a certain task, for the completion of which the agent is conferred a wide margin of independence or freedom of action.

An agency contract differs from an employment contract in which, in general, it is the agent (the employee, in a nutshell) to whom responsibility falls for coordinating, directing and controlling the work commissioned by the principal. To conceptualize the firm as a set of heterogeneous groups: shareholders, directors, creditors, suppliers, clients, etc., implies the existence of a set of different contracts. The costs arise in the firm as a consequence of the conflicts that emerge between these groups with opposing interests. Under this proposal, the firm is organized as a set of principal-agent relations. So, its design should tend to minimize the agency costs that cover all the contractual costs, frequently referred to as transaction costs, costs of moral hazard and information costs, which are incurred with the object of reducing deviations in the behaviour of the agent with regard to the interests of the principal.

It should be recognized that the contracting parties support the agency costs associated with their interaction, and have incentives to enter into contracts that, in so far as possible, reduce those costs. Specifically, the contracting parties benefit from foreseeing actions to negotiate and to sign contracts that facilitate the desired actions. Incentives are thereby generated to enter into contracts and to create institutions with fewer agency costs (Smith 1987).

Agency problems arise from conflicts of interests that are commonly found in most cooperative ventures, both whether they are or are not done in the hierarchical manner implicit in the analogy of principal-agent. This opening of agency costs to all cooperative relations has important implications for agency theory, because when the difference between principal and agent is removed, the distinction between the supervisory costs and finance costs are also lost. In this way, total agency costs are the costs occasioned by actions that aim to reduce residual loss plus the opportunity costs.

The paradigm of agency theory has its immediate antecedents in the literature developed on the basis of the separation between ownership and control. According to Levithal (1988), this current of thought may be understood as the neoclassical response to the questions proposed by Barnard (1938) and March and Simon (1958) on the behaviour of an organizations formed by agents with their own self-interest and objectives in conflict, in a world with incomplete information. It implies that that this theory is based on two basic suppositions; the existence of uncertainty and conflict of objectives. Uncertainty opens the door to opportunistic behaviour and a conflict of objectives between the principal and the agent. Adding the existence of information symmetries, the need to establish incentive systems is proposed.

There is the risk in all agency relations of a deviation of behaviour of the agent with regard to the interests of the principal. Certainly, every time that authority is delegated in the firm and the behaviour of the agent may not be directly supervised, the problem arises of how to resolve this situation in an efficient way. It may therefore be affirmed, in the words of Rumelt et al. (1991) that agency theory deals with the design of incentives and the assignation of decision rights between individuals with opposing interests.

Jensen (1983) established two approaches in the development of agency theory, which he termed Positive Agency Theory and the Principal Agent Theory. Both approaches are directed towards the design of efficient contracts in a Paretian sense. But both approaches also diverge into various dimensions.

Positive agency theory has an empirical orientation. It seeks to identify the situations in which it is likely that the agent and the principal have contradictory objectives and, subsequently, in the description of the governance mechanisms that limited the opportunistic and the egoistical behaviour of the agent resolving in this way the agency problems.

The Principal Agent Theory is markedly mathematical and has an acute normative character. It centres on the optimum design of contracts in accordance with various hypotheses on the preferences of agents and asymmetrical information, being in this sense a worthy effort to extend the optimizing neoclassic model to the economy of organizations.

3.5 Evolutionist Theory

At the heart of the institutionalist current, Hodgson (1998a, b) presented the economic analysis of the firm in terms of resources and capabilities. It considers whether Adam Smith and Karl Marx may be considered as precursors of this analysis; Knight (1921), Penrose (1959) and Richardson (1972) are those who have truly developed the concept of capabilities.

The contractual approach is interested in the transactions between given individuals. It pays less attention to production and technology, as well as to questions of accumulation and growth. In short, this analysis is essentially static, which means it can treat neither dynamic efficiency nor perspectives in the long-term. In particular, the heterogeneity of the behaviours and the performance of firms mean those realities are not apprehended.

In contrast, the approach in terms of capabilities grapples with those topics directly, thereby providing, according to Krafft and Maupertuis (1996), a richer theory of the firm and of institutions. In this analysis, the appearance, the structure and the limits of the firm are explained by the presence of individual but also collective capabilities, which are in any case, preserved and reinforced by this organization.

The resources school underlines the differences in resources and capabilities that the firm possesses and the importance of this fact to explain the differences in the results over time. This approach seeks to explain the processes of dynamic change.

Resources are a changing stock of available factors that the firm possesses or controls. They may be either tangible or intangible. The capabilities of the firm are knowledge and skills that arise from the collective learning of the organization, as a consequence of the creation of organizational routines that are developed by the exchange of information between the members of the firm.

It should be highlighted that evolutionist developments that seek to give a theoretical basis to the economic behaviour of the firm frequently recur to contractualist arguments to understand the organization of productive activities. It may be noted that the analysis of dynamic capabilities is oriented towards an attempt to summarize, or at least towards conciliation between the transaction cost economy and the economy of capabilities.

In their seminal work, Nelson and Winter (1982) characterized the firm as a set of capabilities, some of which were intangible, subjected to a process of routinization. In other words, the firm is a hierarchy of organizational routines. The conversion of organizational activities into routines constitutes the principal form of storing its specific operative knowledge; routines as organizational memory. This organizational memory helps to reconsider decisions that are relatively satisfactory in the presence of complex decisions, limited satisfaction or rationality, and to coordinate the respective actions in the absence of perfect communication.

According to Nelson (from Krafft and Maupertuis 1996): "The idea of capabilities that are changing over time, whether within the firm, whether because of exchanges between firms, is fundamental and can contribute a lot to understanding the nature of the firm and its evolving strategies. A certain number of articles have centred on the attempt by firms to accumulate capabilities that are not easily imitable by rival firms. From my point of view, the truly interesting question is not linked to the creation of knowledge that is difficult to imitate. It is rather to study how firms are able to obtain benefits from their knowledge and capabilities".

4 Concluding Remarks: The New Industrial Organization

The evolution of thought relating to the Organization and Management Sciences of has been closely linked to changes in the environment that surrounds the firm. At first, the key concept is management to achieve coordination, in response to the need for integration, of both functional activities and divisions.

Subsequently, as from the 1970s, and in reaction to the acceleration of the process of change in the environment, the emphasis of management was progressively shifting from coordination to the definition and the implementation of strategies, focusing business management not only on internal administration and the type of competition. The work of Learned et al. (1965) pioneered progress in this field.

The firm arose as an economic organization to produce goods and services and as an alternative resource assignation to the market. "A firm will have to grow until the organizational costs of an extra transaction within the firm are equal to the costs involved in completing the transaction in the open market, or the organization costs that relate to the entrepreneur." (Coase 1937). In other words, until the market costs of that additional transaction are equal to the agency costs. It is therefore a problem of equilibrium between business efficiency and where the market ends is the result of a comparative dynamic continuum between transaction and production costs.

As previously stated transaction, cost arise from information, coordination (negotiation), follow up and guarantees to transaction completion. Its origin lies in human factors: limited rationality and opportunism. In contextual factors, repetition and sequencing of stages in the negotiation, relative information asymmetries and the order of intervention in the negotiation. Its amount depends on the specificity of the assets involved, on the frequency of the transaction, its complexity, its uncertainty and possible relations with other transactions.

Hernández (1997) states that transaction costs require recognition and guarantees of the ownership rights, they underline the importance of institutions, and bring to the foreground the art of integrating technology and organization—Engineering of the Organization—, as keys to understand the generation of wealth. The award of the Nobel Prize to R. Coase in 1991 helped his influence to move fast the field of the economy of the firm, forcing convergence into what we may call the New Industrial Organization, of the Economy and the Law, Organizational Engineering and Industrial Organization.

With this approach, it is possible to analyze the different facts that determine the behaviour of organizations and of markets. In other words, how the markets and firms structure themselves, how the sellers, the consumers, the workers and the intermediaries behave in those markets, and how those markets function from the point of view of prices, costs, product quality, innovation, risk distribution and other operational indicators. Without forgetting to mention aspects linked to the organization of management and production in the firm.

In our understanding, the object of Industrial Organization within Economics is represented well by J. Tirole and his work The Theory of Industrial Organization (1989). Basically, it centres on how equilibrium is achieved with a set of alternative economic criteria to the classical competitive market with complete contracts and non-symmetrical information. In fact, equilibrium may be inefficient when comparing it with the earlier norm. Prices can exceed the marginal cost and the quality of the product may be very high or very low. It may have many or few products, etc.

It is a question of understanding how economies of scale and hidden costs, asymmetrical information, product differentiation, and other basic economic characteristics of a market, combined with different behavioural hypotheses, affect the resulting equilibrium of imperfect competition that is associated with the operation of the market. In other words, it explores the extensive terrain of imperfect competition that exists between the simple models of perfect competition and the pure classic monopoly.

Few markets are pure competition or pure monopolies. Hence, Industrial Organizations plays a very important role, when providing a refined characterization of what competition means in imperfect markets. In addition, we may create a coherent structure to analyze how changes in the institutional environment (common lays, administrative regulation, etc.) affect the structure and the operation of the market and the behaviour of the firm.

What is more, the Industrial Organization also provides links with Organizational Engineering (governance structure) because the existence of imperfect competition can mean that vertical integration or non-standard contracts are attractive governance alternatives, in order to respond to those market imperfections or to increase the power of the market. However, it largely ignores the institutional environment and the governance structure.

The integration of the institutional environment implies the interrelations between its attributes and the organization and the operation of the markets. It is fundamentally interested in the evolution and the role of formal and informal institutions that govern ownership rights in the market, the nature and impact of regulation on its operation, and its organization. The most relevant aspect is the evolution of ownership rights.

Finally, Organizational Engineering (design, planning, and control of productive systems) may be mentioned, as a generator of wealth and improved coordination; for example, the adoption of assisted productive systems, such as JIT, initially developed by Toyota. It is a system of communication and close coordination between the successive phases of a process, provoking an increase in the reliability of the system. The improvement of the process simultaneously provokes an improvement in product quality. It is, in this way, possible to compete against firms with large economies of scale, through greater flexibility linked to shorter response times.

The adoption of these techniques, which are better forms of coordinating production, by other industries together with the new approaches to management from total quality and virtual organizations, have meant that those economies that have adopted these methods have increased their productivity and reduced the volume of their inventories.

What is it that makes the New Industrial Organization different from the wide and diverse set of methodological approaches for the analysis of firms and the market?

We consider the New Industrial Organization as something the flows from the paradigm of the Industrial Organization, expanding towards a richer and more complete specification of the industrial environment and the transactional variables that characterize the organization of firms and markets. In addition, feedback and interactions between the institutional setting and the structure are taken into account, as well as the behaviour and operation of firms.

From the standpoint of exchange costs and production costs, we are interested in the transaction costs, considering goods and services as transactions with their attributes, in other words, the exchange of rights.

The institutional environment emphasizes ownership rights, legal institutions, and the clients and norms that play a fundamental role to determine how the markets are organized and, more importantly, how they behave. As Williamson (1993) observed, the differences between the governance structure and the institutional environment are probably more important to explain the international and inter-temporal differences in organizational agreements.

Economic agents seek their own interests, but operate in a world of limited rationality. Information is costly, is asymmetrically distributed and the contracts are incomplete. The institutional and contractual agreements arise to reduce the direct and indirect costs of acquiring inputs, manufacturing products, and selling products and services.

Relevant transaction costs include the costs of carrying out the transaction and, more importantly, the costs of contractual failure (suspension and opportunism), and the costs incurred when dealing with rent-seeking and defending oneself from it.

The essence of the governance structure is, therefore, how to organize, control and consume transactions between economic agents, considering the productive function in its widest sense. At all times under the prism of evolutionist thought, industrial dynamics and a changing environment.

So, the New Industrial Organization (NOI) has three dimensions (Fig. 1), which in the terminology of Hernández (1997) are as follows:

- Vector of opportunities and market threats: basic market conditions, number of agents, competitive interactions and strategic behaviours, information asymmetries and market power.

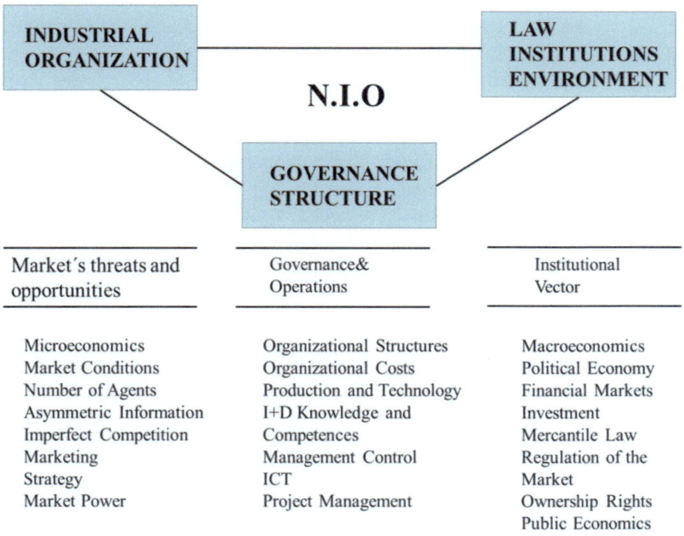

Fig. 1 The dimensions of the New Industrial Organization (NIO)

- Institutional Vector (of economy and law): ownership rights, administrative regulation of those rights, policy of industrial "promotion" and patent protection, etc.
- Vector of governance structures of the firm: technology and production, coordination and motivation, knowledge generation and networking coordination with stakeholders and other firms to maintain the firm as an open system. According to Joskow (1995), it is a matter of efficient coupling of resources and capabilities of the firm with the opportunities of the market and the institutions. Being in the XXI century we should add: In a global world thanks to the advances of the information and communication technologies.

These dimensions can give us a map of the Management and Organization Sciences, and help designing the Management Engineering curricula and the range of specific skills and competences demanded by different institutions.

References

Alchian, A. A. (1950). Uncertainty, evolution, and economic theory. *Journal of Political Economy, 58,* 211–221.

Alchian, A. A., & Demsetz, H. (1972). Production, information costs and economic organization. *American Economic Review, 62,* 777–795.

Alchian, A. A., & Woodward, S. (1988). The firm is dead; long live the firm. A review of Oliver E. Williamson's. The economic institutions of capitalism. *Journal of Economic Literature, 26,* 65–79.

Barnard, C. I. (1938). *The functions of the executive*. Cambridge: Harvard University Press.

Baumol, W. J. (1961). *Economic theory and operations analysis*. Englewood Cliffs: Prentice Hall.

Caballero, R., & Lyons, R. (1990). Internal versus external economies in european industry. *European Economic Review, 34,* 805–826.

Cantillon R (1775). Essai sur la nature du commerce en général. Fletcher Gyles, London.

Casson, M. (1982). *The entrepreneur -an economic theory*. Oxford: Martin Robertson.

Casson, M. (1990). *Entrepreneurship*. Aldershot: Elgar.

Chandler, A. D. (1977). *The visible hand: The managerial revolution in american business*. Cambridge, MA: Harvard Univ Press.

Coase, R. (1937). The nature of the firm. *Economica, 4,* 386–405.

Coase, R. (1960). The problem of social cost. *Journal of Law and Economics, 3,* 1–44.

Commons, J. R. (1934). *Institutional economics*. Madison: Univ Wisconsin Press.

Demsetz, H. (1967). Toward a theory of property rights. *American Economic Review, 57*(2), 347–359.

Douma, S., & Schreuder, H. (1998). *Economic approaches to organizations*. London: Prentice Hall.

Frank, R. H. (1991). *Microeconomics and behavior*. New York: McGraw-Hill.

Friedman, M. (1991). Old wine in new bottles. *The Economic Journal, 101*(404), 33–40.

Furubotn, E. G., & Pejovich, S. (1972). Property rights and economic theory: A survey of recent literature. *Journal of Economic Literature, 10*(4), 1137–1162.

Furubotn, E. G., & Pejovich, S. (1974). *The economics of property rights*. Cambridge, MA: Ballinger.

Gigerenzer, G., Czerlinski, J., & Martignon, L. (2002). How good are fast and Frugal heuristics? In T. Gilovich, D. Griffin, & D. Kahneman (Eds.), *Heuristics and biases: The psychology of intuitive judgement*. Cambridge: Cambridge University Press.
Gordon, R. J. (1993). Why the principles course needs comparative macro and micro. *American Economic Review, 83*(2), 17–22.
Hayek, F. A. (1988). *The fatal concept: the errors of socialism*. London: Routledge.
Hernández C (1997). Crecimiento Económico, Progreso Institucional y Factor "e". España en la Unión Monetaria. Lección de Apertura del Curso Académico 1997–98. UVA, Valladolid.
Hirshleifer, J. (1956). On the economics of transfer pricing. *Journal of Business, 29*(3), 172–184.
Hodgson, G. M. (1998a). Evolutionary and competence based theories of the firm. *Journal of Economic Studies, 25*(1), 25–56.
Hodgson, G. M. (1998b). The approach of institutional economics. *Journal of Economic Literature, 36,* 166–192.
Jensen, M. C., & Meckling, W. H. (1976). Theory of firm: Managerial behavior, agency costs and ownership structure. *Journal of Financial Economics, 3*(4), 305–360.
Jensen, M. C. (1983). Organization theory and methodology. *Accounting Review, 58*(2), 319–339.
Joskow, P. L. (1995). The new institutional economics: alternative approaches. *Journal of Institutional and Theoretical Economics, 151*(1), 248–259.
Kahneman, D. (1994). New challenges to rationality assumption. *Journal of Institutional and Theoretical Economics, 150*(1), 18–36.
Kahneman, D., & Tversky, A. (1979). Prospect theory: An analysis of decision under risk. *Econometrica, 47*(2), 263–292.
Kirzner, I. (1982). The theory of entrepreneurship in economic growth. In C. A. Kent, D. L. Sexton, & K. H. Vesper (Eds.), *Encyclopedia of entrepreneurship*. Englewood Cliffs: Prentice Hall.
Kirzner, I. (1997). Entrepreneurial discovery and the competitive market process: An Austrian approach. *Journal of Economic Literature, 35*(1), 60–85.
Knight, F. H. (1921). *Risk, uncertainty and profit*. Boston: Schaffner & Marx.
Knudsen, C. (1995). Theories of the firm, strategic management, and leadership. In C. A. Montgomery (Ed.), *Resource-based and evolutionary theories of the firm: Towards a synthesis* (pp. 179–217). New York: Springer.
Krafft, J., & Maupertuis, M. A. (1996). Institutionnalisme, évolutionnisme et Compétences des Firmes: Le point de vue de Richard Nelson. *Revue d'Economie Industrielle, 78,* 68–80.
Learned, E., Christensen, R., Andrews, K., & Guth, W. (1965). *Business policy: Text and cases*. Homewood: Irwin.
Leibenstein, H. (1987). *Inside the firm: The inefficiencies of hierarchy*. Cambridge, MA: Harvard Univ Press.
Levithal, D. (1988). A survey of agency models of organizations. *Journal of Economic Behavior & Organization, 9*(2), 153–185.
Lipsey, G. R. (1963). *An introduction to positive economics*. London: Weidenfeld & Nicolson.
López-Paredes, A., Hernández, C., & Pajares, J. (2002). Towards a new experimental socio-economics: Complex behaviour in bargaining. *Journal of Behavioral and Experimental Economics (formerly The Journal of Socio-Economics), 31*(4), 423–429.
Lucas, R. (1993). Making a miracle. *Econometrica, 61*(2), 251–272.
Llewellyn, K. N. (1931). What price contract? An essay in perspective. *Yale Law Journal, 40*(5), 704–751.
March, J. G., & Simon, H. (1958). *Organizations*. New York: John Wiley.
Milgrom, P., & Roberts, J. (1992). *Economics, organization, and management*. Englewood Cliffs: Prentice Hall.
Naylor, T., & Vernon, J. (1969). *Microeconomics and decision models of the Firm*. Brace and World, New York: Harcourt.
Nelson, R., & Winter, S. (1982). *An evolutionary theory of economic change*. Cambridge, MA: Belknap Press.
Penrose, E. (1959). *The theory of the growth of the firm*. Oxford: Blackwell.

Pindyck, R. S., & Rubinfeld, D. L. (1998). *Microeconomics*. Upper Saddle River: Prentice Hall.

Posada, M., & López-Paredes, A. (2008). How to choose the bidding strategy in continuous double auctions: Imitation versus take-the-best heuristics. *Journal of Artificial Societies and Social Simulation, 11*(1), 6 http://jasss.soc.surrey.ac.uk/11/1/6.html

Putterman, L., & Kroszner, R. (1986). *The economic nature of the firm: A reader*. New York: Cambridge Univ Press.

Richardson, G. (1972). The organization of industry. *Economic Journal, 82,* 883–896.

Romer, P. (1990). Endogenous technological change. *Journal of Political Economy, 98*(5), S71–S102.

Rumelt, R., Schendel, D., & Teece, D. (1991). Strategic management and economics. *Strategic Management Journal, 12*(S2), 5–30.

Salanié, B. (1997). *The economics of contracts*. Cambridge, MA: MIT Press.

Schumpeter, J. A. (1954). *History of economic analysis*. London: Allen & Unwin.

Simon, H. (1947). *Administrative behavior*. New York: McMillan.

Simon, H. (1997). *Models of bounded rationality* (Vol. 3). Cambridge, MA: MIT Press.

Smith C. W. (1987). Agency costs. In: J. Eatwell, M. Milgate, P. Newman (eds) *The new palgrave: A dictionary of economic theory and doctrine* vol. 1 (pp. 39–40). London: McMillan.

Solow, R. (1956). A contribution to the theory of economic growth. *Quarterly Journal of Economic, 70*(1), 65–69.

Tirole, J. (1989). *The theory of industrial organization*. Cambridge: MIT Press.

Williamson, O. E. (1975). *Markets and hierarchies: Analysis and antitrust implications*. New York: Free Press.

Williamson, O. E. (1985). *The economic institutions of capitalism: Firms, markets and relational contracting*. New York: Free Press.

Williamson, O. E. (1986). *Economic organization: Firms, markets and policy control*. Brighton: Wheatsheaf.

Williamson, O. E. (1993). Transaction cost economics and organization theory. *Industrial and Corporate Change, 2*(2), 107–156.

Part II
The Governance System of the Firm

Chapter 5
The Application of the Viable System Model to Enhance Organizational Resilience

Cristina Ruiz-Martin, José Manuel Pérez Rios, Gabiel Wainer, Javier Pajares, Cesáreo Hernández and Adolfo López Paredes

Abstract The interest in how to build resilient organizations is increasing in the last two decades. However, there is no formal and accepted framework yet. In this paper, we argue that the application of the principles of the Viable System Model (VSM) improves organizational resilience. We also argue that the VSM constitutes a valid framework to design resilient organizations.

Keywords Organizational resilience · Viable system model · Organization design

1 Introduction

The study of resilience is gaining attention in the research agenda. A recent search in Scopus shows that there are more than 71.000 documents talking about resilience, resilient or resiliency. More than 62.000 of them published since the year 2.000. The number of papers related to resilience is increasing every year.

C. Ruiz-Martin (✉) · J.M. Pérez Rios · G. Wainer · J. Pajares · C. Hernández · A. López Paredes
INSISOC, Universidad de Valladolid, Paseo del Cauce 59, 47011 Valladolid, Spain
e-mail: cruiz@eii.uva.es

J.M. Pérez Rios
e-mail: rios@uva.es

G. Wainer
e-mail: gwainer@sce.carleton.ca

J. Pajares
e-mail: pajares@eii.uva.es

C. Hernández
e-mail: cesareo@eii.uva.es

A. López Paredes
e-mail: aparedes@eii.uva.es

C. Ruiz-Martin · G. Wainer
Department of Systems and Computer Engineering, Carleton University,
1125 Colonel by Drive, 3216 V-Sim, Ottawa, ON K1S 5B6, Canada

C. Hernández (ed.), *Advances in Management Engineering*, Lecture Notes in Management and Industrial Engineering, DOI 10.1007/978-3-319-55889-9_5

Resilience it is studied in different research fields, including ecology, psychology, disaster management, organization management, sociology, engineering, etc. This is probably why there is no common and widely accepted and unified definition of resilience. Even within the same area, different definitions coexist (Bergström et al. 2015). Despite those areas seem to be isolated and independent from each other, they are not.

Rose (2004) focuses on economic resilience and proposes that resilience takes place at three different levels: microeconomic, mesoeconomic and macroeconomic. At microeconomic level, we care about the resilience of the individual behaviour of firms, households and organizations. At mesoeconomic level, we focus on the resilience of an economic sector, individual market or cooperative group. Finally, at the macroeconomic level, we combine all individual units and markets. At macroeconomic level, the whole is not just the sum of the parts due to interactive effects of economy. Following the perspective presented in Rose (2004), we suggest that the different research areas studying resilience can be linked together. All of them study resilience at one of the above-mentioned levels. For example, to have a resilient organization we need to have resilient individuals (Mallak 1998b), among other requisites.

Our study focuses on resilience at the microeconomic level, specifically at the level of organizations. At this level, several works have proposed principles that we should follow to develop resilient organizations and the characteristics a resilient organization should have.

For example, Mallak (1998a, b) propose seven principles to create a resilient organization: perceive experiences constructively, perform positive adaptive behaviors, ensure adequate external resources, expand decision-making boundaries, practice bricolage, develop tolerance for uncertainty and build virtual role systems.

Similarly, Coutu (2002) states that a resilient organization has to face down reality, search for meaning and continually improvise. Dervitsiotis (2004) proposes that a resilient organization has the characteristics of living systems: receptivity from early warning systems, flexibility and capacity of creativity and innovation. We have reviewed over 200 papers and we have found that these approaches lack a formal framework to create resilient organizations.

The Viable System Model, here after VSM, (Beer 1981) is a scientific framework based on organizational cybernetics applied to the design and study of organizations and its processes (Pérez Ríos 2012). In the management field, the application of VSM is taking more attention.

A Viable System is a system organized in a way that it is able to survive despite changes in its environment. Preis (2014) has already proposed a framework for resilient management based on the principles of organizational cybernetics. However, he only takes into account one of the principles of the VSM: the recursive character.

Considering the definition of viable system and the aim of resilience (aligned with Dervitsiotis op.cit) we propose that the application of the VSM principles to organizations improves its resilience.

The rest of the paper is organized as followed. In Sect. 2, we explain the application of the VSM to organizations. In Sect. 3, we briefly review the concept of organizational resilience, the factors that contribute to organizational resilience and how it is measured. In Sect. 4, we explain how the VSM is a valid and an appropriated framework to design resilient organizations. Finally, in Sect. 5, we present the conclusions of this work.

2 The Viable System Model. Application to Organizations

The challenge that leaders and managers in organizations face in the current turbulent environment is formidable. The complex environment in which they act demands that managers have access to decision-making tools commensurate with the complexity which they must face (Schwaninger and Pérez Ríos 2008). In relation to this issue of the capacity for handling complexity, it has been pointed out that the quality of decisions made by managers is limited by the quality of the models they use for the systems they try to govern. If we are concerned with the viability of an organization [understood as system—see Beer (1989)], meaning with this term the capacity of a system to maintain a separate existence, (i.e. to survive regardless of changes in its environment), then we can apply an organizational cybernetic approach, in particular the Beer's Viable System Model (VSM). According to the VSM a viable organization must have the capacities of self-regulation, learning, adaptation, and evolution.

In his Viable System Model (VSM), Beer (1981, 1985) establishes the necessary and sufficient conditions for the viability of an organization. These are related to the existence of a set of functional systems (Beer identified them as System 1, 2, 3/3*, 4 and 5) in an organization and a set of relationships among these functional systems and the environment. These systems and the relations among them are represented in Fig. 1.

According to Beer, all viable systems contain viable systems and are themselves contained in viable systems. The most important aspect of this recursive conception of viable systems is that, no matter which place they occupy within the chain of systems, they must always contain the five functional systems that determine viability, in order to be viable.

System 1 is responsible for producing and delivering the goods or services which the organization produce. In the example shown in Fig. 1, System 1 is made up of three elemental operational units (Op. Unit 1, 2 and 3) which can be divisions of a company, suborganizations, etc. The main role of *System 2* is to guarantee a harmonic functioning of the organizational units, which compose system 1. *System 3* is responsible for optimizing the functioning of the whole set of system 1, made up of the different operational units. We can say that it is responsible for the "here and now" of the organization. The main responsibility of *System 4* is to monitor the

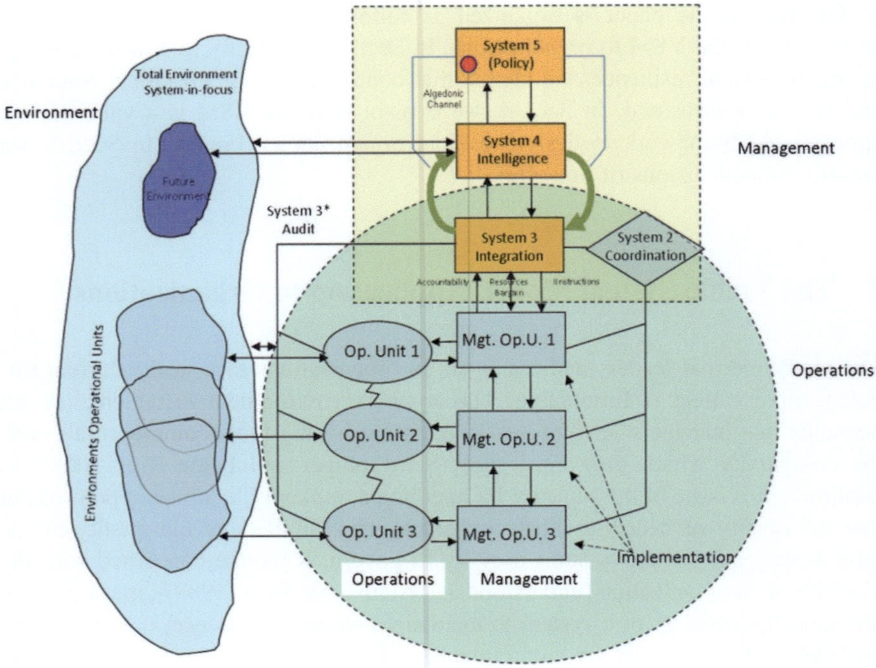

Fig. 1 Viable system model, adapted from Beer (1981) (Pérez Ríos 2012)

environment of the organization. It takes care of the "outside and then" of the organization, with the aim of maintaining it always prepared to change. *System 5* takes care of the normative decisions and is responsible for defining the ethos, the vision and the identity of the organization.

Based on Organizational Cybernetics (OC) and, in particular, the VSM's conceptual elements, Pérez Ríos (2010) introduced a systemic methodological framework to help design or diagnose systems in view of their viability. The process to apply it is structured in four main steps as we show in Fig. 2.

The *first step* is to identify the identity and the purpose of the organization. In this process, we will try to assess what the organization is (and also, what the organization is not) and what it is, or should be, its purpose.

In a *second step,* we see how the organization faces the total environment complexity (variety) by means of creating a vertical structure made up of sub-organizations where each of them will be in charge of the different sub-environments in which the total environment is also divided.

In a *third step*, we should go through each of those vertical levels and get into them to check that all the necessary and sufficient elements for viability, which OC and the VSM identifies are adequately represented in all the organizations, sub-organizations, sub-sub-organizations, etc. in which we have unfolded the initial organization.

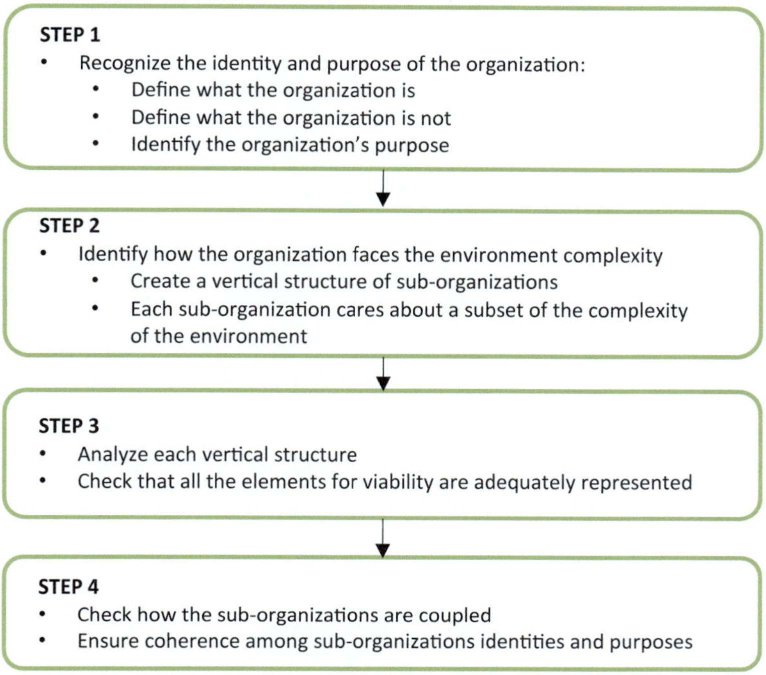

Fig. 2 Framework to help design and diagnose systems in view of their viability

The *fourth* and last step would be to check the degree of coupling of all organizations, sub-organizations etc. at all recursion levels, from the point of view of the coherence among their respective identities and purposes.

We want to highlight that any shortage in any of these five systems or functions due to absence, to malfunction or to deficient design of the communication channels that connect them carries pathologies in the organization. These pathologies cause that the organization does not work properly or even disappear, at least as an independent entity.

The variety of pathologies which most frequently appear in organizations have been analysed and classified comprehensively by Pérez Ríos (2012) into three main groups:

a. Structural Pathologies are related to the organization structural design and how it copes with the total environmental complexity and creates the necessary sub-organizations.
b. Functional Pathologies are those related to the adequacy of organizations (at all recursion levels) to the prescription made by the VSM about functions and subsystems and their relationships.
c. Information pathologies are related to information systems and communication channels.

Both a comprehensive definition of the VSM and the pathologies that organizations face can be found in (Beer 1989; Schwaninger 2009; Hetzler 2008; Pérez Ríos 2008, 2012).

3 Organizational Resilience. A Briefly Review

Reviewing over 50 definitions, we have found that most of the authors understand organizational resilience as an **ability** such as (Mallak 1998a), **capacity** such as (Manyena 2006) or **capability** such as (Annarelli and Nonino 2016) to deal with internal or external changes, risks or jolts. For example, Mallak (1998a) define resilience as "*the ability of an individual or organization to expeditiously design and implement effective strategies and actions matched to the immediate situation, while enduring minimal stress*". Manyena (2006) define disaster resilience as an "*intrinsic capacity of a system, community or society predisposed to a shock or stress to adapt and survive by changing its non-essential attributes and rebuilding itself*".

Based on a recent literature review, Annarelli and Nonino (2016) provides a more integrative definition of organizational resilience. They define organizational resilience as "*the organization's capability to face disruptions and unexpected events in advance, thanks to the strategic awareness and a linked operational management of internal and external shocks. The resilience is static, when founded on preparedness and preventive measures to minimize threats probability and to reduce any impact that may occur. And it is dynamic, when founded on the ability of managing disruptions and unexpected events to shorten unfavorable aftermaths and maximize the organization's speed of recovery to the original or to a new more desirable state*". This definition captures some key concepts of the definitions provided in the literature such as the capability to face disruptions, being prepared for disruptions, prevent them and shortening the recovery time and the consequences to achieve the original state or a more desirable one. However, this definition does not explicitly take into account that changes and disruptions can be opportunities (Bhamidipaty et al. 2007; Ates and Bititci 2011), the learning process in organizations (Stewart and O'Donnell 2007; Chand and Loosemore 2012), the importance of social interactions (Powley 2009), and the capacity to innovate (Robb 2000; Reinmoeller and Van Baardwijk 2005) and reinvent (Hamel and Valikangas 2003; Mafabi et al. 2015).

The factors or characteristics that contribute to create resilient organizations have also been largely studied based on both theoretical and empirical works. We have found a great variety among the factors and mechanisms that contribute to enhance organizational resilience. For example, Riolli and Savicki (2003) propose that organizational resilience is based on resilient individuals. However, they also acknowledge that having resilient individuals do not guarantee organizational resilience.

Fiksel (2003) propose that resilient systems, such as enterprises, have the following characteristics: diversity, efficiency, adaptability and cohesion. Jackson (2007) considers the following properties: adaptability, agility and robustness. McManus et al. (2008) propose to enhance resilience improving situation awareness, the management of keystone vulnerabilities and the adaptive capacity. Van Trijp et al. (2012) add taking into account quality to the previous factors. Berliet (2009) suggests three pillars to enhance resilience: enterprise risk management, value-based management and management by objectives. Burnard and Bhamra (2011) remark the importance of enhance monitoring, adaptive capacity, self-assessment of vulnerabilities, flexibility and organizational learning. Sanchis and Poler (2013) identify vulnerability, adaptive capacity and recover ability as the components of resilience.

Other authors performed empirical studies to identify the factors that contribute to resilience. For example, Crichton et al. (2009) found common lessons learned to improve resilience through the study of different incidents in the UK. Powley (2009), through the study of an university shooting, found three mechanisms that activate resilience: liminal suspension, compassionate witnessing and relational redundancy. Beermann (2011) concluded that the combination of mitigation and adaptation strategies helps to create more robust and resilient strategies after studying different organizations in the German food industry. We want to acknowledge that not all the factors or characteristics identified in these empirical works may be applicable to all types of organizations or business sectors.

However, though a literature review, we found some common and repeated characteristics or factors to take into account to enhance resilience. These factors include building situation awareness (Coutu 2002; McManus et al. 2008), managing organization's vulnerabilities (Erol et al. 2010), having resources (Orchiston et al. 2016; Mallak 1998b), having improvisation capacity (Kendra and Wachtendorf 2002; Coutu 2002; Mallak 1997), being able to anticipate to events (Hardy 2014; Apneseth et al. 2013), being agile (Gibson and Tarrant 2010; Thomas et al. 2016), having learning capacity (Burnard and Bhamra 2011; Robb 2000), collaboration (Andrés and Poler 2013; Winston 2014), having resilient individuals (Mallak 1997; Riolli and Savicki 2003) and being flexible (Kendra and Wachtendorf 2002; Proper and Pienaar 2011) and redundant (Chopra and Khanna 2014; Winston 2014).

The number of authors that have investigated how to measure organizational resilience is lower than the ones that have identified factors that contribute to enhance resilience. Many authors have proposed to assess resilience evaluating how the organization performs in the different factors or characteristics that contribute to resilience (Horne and Orr 1998; Bhamidipaty et al. 2007; Somers 2009; Sanchis and Poler 2013; Lee et al. 2013; Seville 2009; Whitman et al. 2013). However, they have not reached an agreement. This lack of agreement is mainly based on the lack of agreement about the factors or characteristics that contributes to resilience.

For example, Seville (2009) proposes 23 indicators with a description to evaluate the factors that contribute to resilience: resilience ethos, situation awareness,

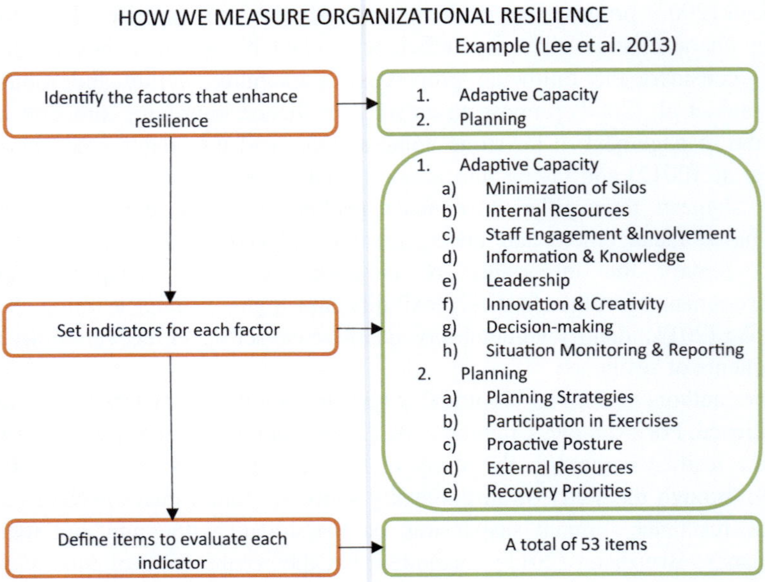

Fig. 3 Example of how to measure resilience following Lee et al. (2013)

management of key stone vulnerabilities and adaptive capacity. These indicators include commitment to resilience, network perspective, informed decision-making or recovery priorities.

Lee et al. (2013) propose a two-factor model to assess organizational resilience (see Fig. 3). They suggest that these factors are adaptive capacity and planning. They evaluate the factors measuring 13 indicators (such as minimization of silos, internal resources or planning strategies) based on 53 items.

For example, they propose to measure the capability and capacity of internal resources based on 3 items. These items are:

- Enough resources to successfully operate during business-as-usual
- The resources during business as usual are manage in a way that can absorb small amounts of unexpected change
- There is less paperwork to have the internal resource available when there is a problem in the organization.

Based on the 53-item resilience assessment tool developed by Lee et al. (2013), Whitman et al. (2013) propose a shorter version of the tool with 13 items of the 53 using just one item for each indicator. The results they obtained with the shorter version are correlated with the ones they got with the 53-item scale. The advantage is that it carries less time to assess resilience, as there are fewer items to be evaluated.

This brief review points out that future research lines should aim at identifying a framework of core enabling factors or characteristics that contribute to enhance resilience and allow evaluating it.

4 The Application of the Viable System Model to Design Resilient Organizations

Organizational Cybernetics applies "communication and control" cybernetic principles to the organizations (Pérez Ríos 2010). We have already explored the VSM and we have identified the concepts that link VSM and organizational resilience.

A review of several definitions of resilience have pointed out that, among other characteristics, resilient organizations have to recover from challenges or disruptive events (i.e. survive) (Sheffi and Rice 2005; Fiksel 2006; Manyena 2006; Stewart and O'Donnell 2007; Hollnagel 2010; Annarelli and Nonino 2016). Therefore, a resilient organization has to be a viable one.

The VSM establishes the necessary and sufficient conditions for the viability of an organization (Beer 1979, 1981, 1985, 1989). Viability is the capacity of an organism to maintain its separate existence (i.e. ability to survive despite changes in the environment). The viability of the organization is related to the existence of a set of systems or functions inside the organization and a set of relations among them and the environment as explained in the previous section. Moreover, according to the VSM a viable organization must have the capacities of self-regulation, learning, adaptation, and evolution.

These capacities, among others, are within the set of factors that contribute to enhance organizational resilience or, within the set of characteristics and properties a resilient organization should have. For example, McManus et al. (2008); Van Trijp et al. (2012) or Jackson (2007) consider adaptability as an attribute that a resilient organization should have.

Learning (Stewart and O'Donnell 2007; Robb 2000; Zhang and Van Luttervelt 2011; Hilton et al. 2012; Alexiou 2014) and evolution (Demmer et al. 2011) are also included among the factors and characteristics of resilient organizations. Other authors (Fiksel 2006; Proper and Pienaar 2011) do not explicitly talk about evolution, but they include grow (which can be understood as evolution) among the characteristics of resilient organizations. Self-regulation, understood as absorbing environment variability, is also included among the characteristics of resilient organizations (Linnenluecke and Griffiths 2010; Jaaron and Backhouse 2014). This explanation is summarized in Fig. 4.

Following this analysis, we conclude that resilient organizations fit the VSM principles. Therefore, the systemic methodological framework introduced by Pérez Ríos (2010) is a valid and appropriate framework to design a resilient organization.

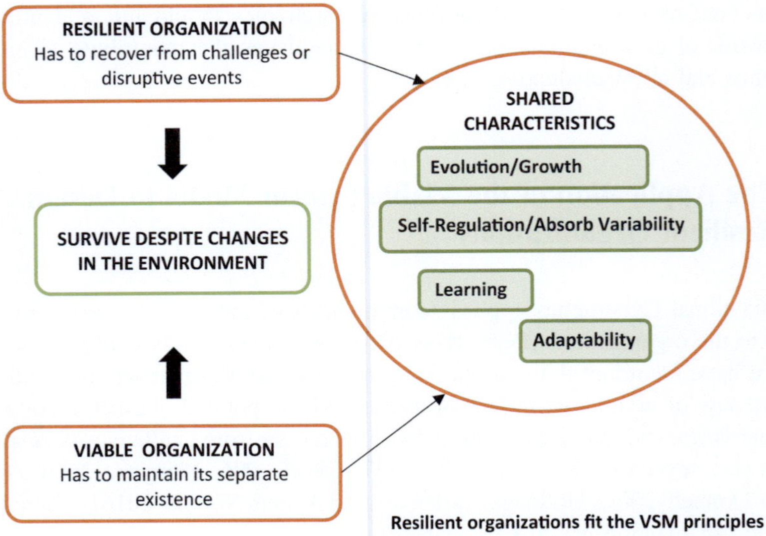

Fig. 4 Shared characteristics between resilient and viable organizations

5 Conclusions

In this work, we have presented the VSM as a framework to design resilient organizations. For this purpose, we have reviewed the principles of the VSM and the framework proposed by Pérez Ríos (2010) to design viable organizations based on the VSM and Organizational Cybernetics. We have also reviewed the concept of organizational resilience. We have reviewed several definitions, the factors that contribute to enhance organizational resilience and how to measure it. We have concluded that despite it is difficult to reach a consensus among the factors that contribute to enhance organizational resilience and how to measure it. Future research lines should tackle this problem.

The VSM establishes the necessary and sufficient conditions for the viability of an organization. The viability of an organization is its ability to survive despite changes in the environment. Taking into account this definition, a resilient organization has to be a viable one. Moreover, according to the VSM, a viable organization must have the capacities of self-regulation, learning, adaptation, and evolution. These capacities are also stated among the factor or characteristics that resilient organizations should have. Therefore, the VSM provides a valid formal framework to design resilient organizations. More specifically, the systemic methodological framework introduced by Pérez Ríos (2010) is a valid and appropriate framework to design a resilient organization.

Acknowledgements This research has been partially supported by University of Valladolid, Banco Santander and NSERC.

References

Alexiou, A. (2014). Taming the waves of adversity: Exploring the multidimensional construct of organizational resilience. In *Managing Emerging Technologies for Socio-Economic Impact* (pp. 340–353).

Andrés, B., & Poler, R. (2013). Enhancing enterprise resilience through enterprise collaboration. In *IFAC Proceedings Volumes (IFAC-PapersOnline)* (pp. 688–693).

Annarelli, A., & Nonino, F. (2016). Strategic and operational management of organizational resilience: Current state of research and future directions. *Omega, 62,* 1–18.

Apneseth, K., Wahl, A., & Hollnagel, E. (2013). Measuring resilience in integrated planning. In *Oil and gas, technology and humans: Assessing the human factors of technological change* (pp. 129–145).

Ates, A., & Bititci, U. (2011). Change process: A key enabler for building resilient SMEs. *International Journal of Production Research, 49*(18), 5601–5618.

Beer, S. (1981). *Brain of the firm: The managerial cybernetics of organization.* New York: Wiley.

Beer, S. (1985). *Diagnosing the system for organizations.* Chichester: Wiley.

Beer, S. (1979). *The heart of enterprise.* Chichester: Wiley.

Beer, S. (1989). The viable system model: Its provenance, development, methodology and pathology. In R. Espejo & R. Harnden (Eds.), *The viable system model, interpretations and applications of Stafford Beer's VSM.* Chichester: Wiley.

Beermann, M. (2011). Linking corporate climate adaptation strategies with resilience thinking. *Journal of Cleaner Production, 19*(8), 836–842.

Bergström, J., van Winsen, R., & Henriqson, E. (2015). On the rationale of resilience in the domain of safety: A literature review. *Reliability Engineering & System Safety, 141,* 131–141.

Berliet, J. -P., (2009). Increasing the resilience of insurance companies. *Risk Management* (June, pp. 6–9).

Bhamidipaty, A., Lotlikar, R., & Banavar, G. (2007). RMI: A framework for modeling and evaluating the resiliency maturity of IT service organizations. *IEEE International Conference on Services Computing (SCC 2007)* (pp. 300–307). UT: Salt Lake City.

Burnard, K., & Bhamra, R. (2011). Organisational resilience: Development of a conceptual framework for organisational responses. *International Journal of Production Research, 49*(18), 5581–5599.

Chand, A. M., & Loosemore, M. (2012). Hospital facility resilience: An adaptation framework for extreme weather events. In *Association of Researchers in Construction Management, ARCOM 2012 - Proceedings of the 28th Annual Conference* (pp. 101–110).

Chopra, S. S., & Khanna, V. (2014). Understanding resilience in industrial symbiosis networks: Insights from network analysis. *Journal of Environmental Management, 141,* 86–94.

Coutu, D. L. (2002). How resilience works. *Harvard Business Review, 80*(5), 46–56.

Crichton, M. T., Ramsay, C. G., & Kelly, T. (2009). Enhancing organizational resilience through emergency planning: Learnings from cross-sectoral lessons. *Journal of Contingencies and Crisis Management, 17*(1), 24–37.

Demmer, W. A., Vickery, S. K., & Calantone, R. (2011). Engendering resilience in small-and medium-sized enterprises (SMEs): A case study of Demmer Corporation. *International Journal of Production Research, 49*(18), 5395–5413.

Dervitsiotis, K. N. (2004). Navigating in turbulent environmental conditions for sustainable business excellence. *Total Quality Management & Business Excellence, 15*(5–6), 807–827.

Erol, O., Sauser, B. J., & Mansouri, M. (2010). A framework for investigation into extended enterprise resilience. *Enterprise Information Systems, 4*(2), 111–136.

Fiksel, J. (2003). Designing Resilient, Sustainable Systems. *Environmental Science and Technology, 37*(23), 5330–5339.

Fiksel, J. (2006). Sustainability and resilience: Toward a systems approach. *Sustainability: Science, Practice, & Policy, 2*(2), 14–21.

Gibson, C., & Tarrant, M. (2010). A "Conceptual Models" approach to organisational resilience. *The Australian Journal of Emergency Management, 25*(02), 8–14.
Hamel, G., & Valikangas, L. (2003). The quest for resilience. *Harvard Business Review, 81*(9), 52–65.
Hardy, T. L. (2014). Resilience: A holistic safety approach. In *Proceedings—Annual Reliability and Maintainability Symposium* (pp. 1–6).
Hetzler, S. (2008). Pathological systems. *International Journal of Applied Systemic Studies, 2* (special issue: organizational cybernetics in focus, Pérez Rios, J. and Schwaninger, M. [Guest Editors]).
Hilton, J., Wright, C., & Kiparoglou, V. (2012). Building resilience into systems. In *SysCon 2012–2012 IEEE International Systems Conference, Proceedings* (pp. 638–645).
Hollnagel, E. (2010). *How resilient is your organisation? An introduction to the resilience analysis grid (RAG).* Sustainable transformation: Building a resilient organization.
Horne, J. F., & Orr, J. E. (1998). Assessing behaviors that create resilient organizations. *Employment relations today* (pp. 29–39).
Jaaron, A. A. M., & Backhouse, C. J. (2014). Service organisations resilience through the application of the vanguard method of systems thinking: A case study approach. *International Journal of Production Research, 52*(7), 2026–2041.
Jackson, S. (2007). A multidisciplinary framework for resilence to disasters and disruptions. *Journal of Integrated Design and Process Science, 11*(2), 91–108.
Kendra, J., & Wachtendorf, T. (2002). *Elements of community resilience in the world trade center attack* (pp. 97–103).
Lee, A. V., Vargo, J., & Seville, E. (2013). Developing a tool to measure and compare organizations' resilience. *Natural Hazards Review, 14,* 29–41.
Linnenluecke, M., & Griffiths, A. (2010). Beyond adaptation: Resilience for business in light of climate change and weather extremes. *Business and Society, 49*(3), 477–511.
Mafabi, S., Munene, J. C., & Ahiauzu, A. (2015). Creative climate and organisational resilience: The mediating role of innovation. *International Journal of Organizational Analysis, 23*(4), 564–587.
Mallak, L. A. (1997). How to build a resilient organization. In *Proceedings of the industrial engineering solutions 1997 conference* (pp. 170–177) Miami.
Mallak, L. A. (1998a). Measuring resilience in health care provider organizations. *Health manpower management, 24*(4), 148–152.
Mallak, L. A. (1998b). Putting organisational resilience to work. *Industrial Management, 40*(6), 8–13.
Manyena, S. B. (2006). The concept of resilience revisited. *Disasters, 30*(4), 433–450.
McManus, S., et al. (2008). Facilitated process for improving organizational resilience. *Natural Hazards Review, 9*(2), 81–90.
Orchiston, C., Prayag, G., & Brown, C. (2016). Organizational resilience in the tourism sector. *Annals of Tourism Research, 56,* 145–148.
Pérez Ríos, J. (2008). Aplicación de la Cibernética Organizacional al estudio de la viabilidad de las organizaciones. Patologías organizativas frecuentes (parte 2ª y final). *DYNA, 83*(7), 403–422.
Pérez Ríos, J. (2012). *Design and diagnosis for sustainable organizations: The viable system method.* Heidelberg, New York, Dordrecht, London: Springer.
Pérez Ríos, J. (2010). Models of organizational cybernetics for diagnosis and design. *Kybernetes, 39*(9/10), 1529–1550.
Powley, E. H. (2009). Reclaiming resilience and safety: Resilience activation in the critical period of crisis. *Human Relations, 62*(9), 1289–1326.
Preis, A. (2014). A framework for cybernetic resilience management. In *Proceedings of the 5th international disaster and risk conference: Integrative risk management —The role of science, technology and practice, IDRC Davos 2014* (pp. 584–587).

Proper, J. W., & Pienaar, W. J. (2011). Resilience as an imperative in public transport organisations. *Corporate Ownership and Control, 8*(4D), 373–388.

Reinmoeller, P., & Van Baardwijk, N. (2005). The link between diversity and resilience. *MIT Sloan Management Review, 46*(4), 61.

Riolli, L., & Savicki, V. (2003). Information system organizational resilience. *Omega, 31*(1), 227–233.

Robb, D. (2000). Building Resilient Organizations. *OD Practitioner, 32*(3), 27–32.

Rose, A. (2004). Defining and measuring economic resilience to disasters. *Disaster Prevention and Management: An International Journal, 13*(4), 307–314.

Sanchis, R., & Poler, R. (2013). Definition of a framework to support strategic decisions to improve enterprise resilience. In *IFAC Proceedings Volumes (IFAC-PapersOnline)* (pp. 700–705).

Schwaninger, M. (2009). Intelligent organizations. *Powerful Models for Systemic Management* (2nd ed.), Berlin: Springer.

Schwaninger, M., & Pérez Ríos, J. (2008). System dynamics and cybernetics: a synergetic pair. *System Dynamics Review 24*(2), 145–174.

Seville, E. (2009). *Resilience: Great Concept... But What Does it Mean for Organizations? Tephra* (pp. 9–14).

Sheffi, Y., & Rice Jr., J. B. (2005). A supply chain view of the resilient enterprise. *MIT Sloan Management Review, 47*(1), 41–48, 94.

Somers, S. (2009). Measuring resilience potential: An adaptive strategy for organizational crisis planning. *Journal of Contingencies and Crisis Management, 17*(1), 12–23.

Stewart, J., & O'Donnell, M. (2007). Implementing change in a public agency. *International Journal of Public Sector Management, 20*(3), 239–251.

Thomas, A., et al. (2016). Profiling the resiliency and sustainability of UK manufacturing companies. *Journal of Manufacturing Technology Management, 27*(1), 82–99.

Van Trijp, J. M. P., Ulieru, M., & Van Gelder, P. H. A. J. M. (2012). Quantitative approach of organizational resilience for a Dutch emergency response safety region. In *Advances in safety, reliability and risk management —Proceedings of the European safety and reliability conference, ESREL 2011* (pp. 173–180).

Whitman, Z. R., et al. (2013). Short-form version of the benchmark resilience tool (BRT-53). *Measuring Business Excellence, 17*(3), 3–14.

Winston, A. (2014). Resilience in a hotter world. *Harvard Business Review, 92*(4), 56–64, 132.

Zhang, W. J., & Van Luttervelt, C. A. (2011). Toward a resilient manufacturing system. *CIRP Annals - Manufacturing Technology, 60*(1), 469–472.

Chapter 6
Supply Chain Management: The Systemic Challenge

Borja Ponte, Isabel Fernández, Nazario García, Raúl Pino and David de la Fuente

Abstract Practitioners extensively recognize the supply chain as an overall system with growingly complex interdependences. Nevertheless, several studies have pointed out that managers fail to see the whole system, and hence they do not operate consistently to such recognition, but they usually approach the issue from a local optimization perspective. This kind of reductionist solutions create large inefficiencies, which damage the profitability of the various supply chain members. Under these circumstances, we highlight the deployment of the systemic approach as the great challenge of the 21st-century supply chains. This paper develops this idea. Firstly, we explore the concepts of efficiency, flexibility and resilience as core operational goals for supply chains operating under the current global scene. Secondly, we underscore the systemic approach by defining the basic supply chain archetypes underlain by contrary philosophical approaches. Thirdly, we devise a framework based on three strategic axes—education, orchestration, and methodology—from which the systemic challenge must be tackled.

Keywords Supply chain collaboration · Supply chain efficiency · Supply chain flexibility · Supply chain management · Supply chain resilience

---

I. Fernández · N. García · R. Pino · D. de la Fuente
Department of Business Administration, University of Oviedo Polytechnic School of Engineering, Campus de Viesques s/n, 33204 Gijón, Spain
e-mail: ifq@uniovi.es

N. García
e-mail: ngarciaf@uniovi.es

R. Pino
e-mail: pino@uniovi.es

D. de la Fuente
e-mail: david@uniovi.es

B. Ponte (✉)
Logistics Systems Dynamic Group, Cardiff Business School, Cardiff University, Aberconway Building, Colum Drive, CF10 3EU Cardiff, Wales
e-mail: ponteblancob1@cardiff.ac.uk

C. Hernández (ed.), *Advances in Management Engineering*, Lecture Notes in Management and Industrial Engineering, DOI 10.1007/978-3-319-55889-9_6

1 21st-Century Supply Chains: The Operational Goals

The business context has dramatically evolved throughout the last four decades because of the geopolitical restructuring after the Cold War, the liberalization of capital markets, and the decrease of transport costs, among other reasons. This situation has led to a global business scene, where the complexity and dynamism of the (generic and specific) organizational environment has significantly increased. This fact can be illustrated via the large and rapid increase of the world trade merchandise exports in these four decades (see Fig. 1). Managers have now to address long lead times and cultural and administrative differences, as well as they must deal with increasingly changing fashions and new competitors.

This context has been accentuated in recent years due to the revolution of information and communication technologies and the growth of the emerging markets. This has led to the current perception that competition goes further than the product itself, covering the overall concept of supply chain; which alludes to the interactions between the network of independent firms that are involved in manufacturing a product—or offering a service—and placing it in the hands of the end user (Mentzer et al. 2001). Hence, the supply chain involves raw materials and component suppliers, product assemblers, wholesalers and distribution companies, and retailer merchants.

Therefore, the concept of Supply Chain Management (SCM)—which refers to the design, planning, execution, control and monitoring of the associated activities with the aim of creating net value in the various participants of the supply chain (Fawcett et al. 2014)—has acquired a strategic importance. Built on the

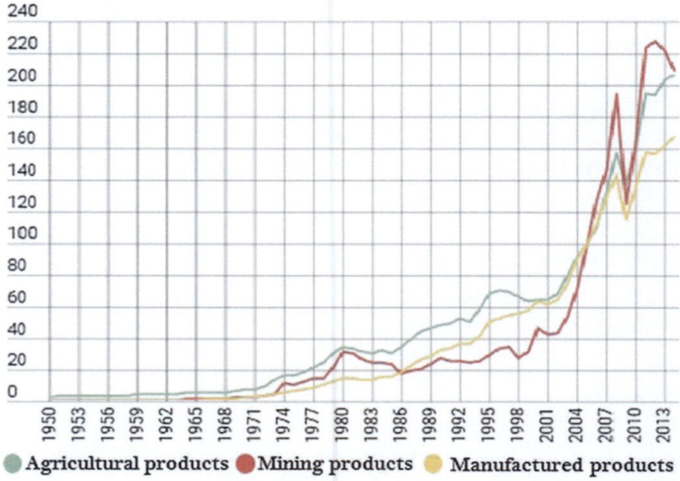

Fig. 1 Evolution in relative terms of world merchandise exports from 1950 to 2014. *Source* World Trade Organization (2016)

aforementioned global business context, 21st-century supply chains must be efficient, flexible, and resilient in order to successfully deal with it.

Efficiency is the clearest concept, as its use is the same in many other fields—that is, the ratio of the obtained yield to the resources used. In terms of SCM, the concept of efficiency refers to the ability of the supply chain to deliver the product to where it is needed, in the required quantity, with the right quality, and at the appropriate time, using in these processes the fewest resources. Being efficient allows the supply chain to improve its current performance, which translates into a profitability increase of its constituent members (Beamon 1999).

Flexibility[1] and resilience can be considered as two sides of the same coin (Husdal 2008). That is, they are different but intrinsically related. Flexibility refers to the inherent capability of the supply chain to alter its current course with the aim of accommodating and successfully adapting to changes in its environment; e.g. redesigning a product, switching its suppliers, or reconfiguring its production and distribution system. A flexible supply chain accommodates disruptions by modifying its direction without affecting its performance. For this reason, flexibility ensures the long-term competiveness of the supply chain (Tachizawa and Thomsen 2007).

Meanwhile, resilience[2] covers the capacity of the supply chain to regain stability after being disturbed. Hence, it refers to the ability to respond when facing harmful disruptions, both uncontrollable (such as political conflicts, natural conflicts, or changes in crude oil price) and partially controllable (like faults in the production system, employee's strikes or suppliers' opportunistic behavior). A resilient supply chain is affected moderately in short term, but it is able to return to the initial conditions. Therefore, being resilient is essential to make survival certain; that is, to ensure the long-term viability of the supply chain (Ponomarov and Holcomb 2009).

These three core concepts for 21st-century supply chains are illustrated in Fig. 2, which is adapted and extended from Husdal (2008). In this graph, the black line symbolizes the supply chain; line thickness represents the time evolution of its profitability. A change of direction in the line refers to significant alterations (required after an associated event) in the course of supply chain. The height of the white triangles indicates the economic value of the resources used. The different kind of disruptions are detailed in the legend of the figure.

---

[1]Similar to other authors (e.g. Asbjørnslett 2009), we understand that flexibility and agility refer to the same idea. Nonetheless, it should be noted that other authors differentiate between them. For instance, according to Goranson (1999), flexibility involves planned adaption to unforeseen but expected external disruptions, while agility involves unplanned adaption to unexpected external circumstances.

[2]Although resilience and robustness are clearly related, Husdal (2010) makes a distinction. While robustness alludes to the ability to endure foreseen and unforeseen changes in the environment without altering the firm's direction, resilience refers to the capability of surviving also foreseen and unforeseen changes (that are generally more severe and/or long-lasting according to this definition).

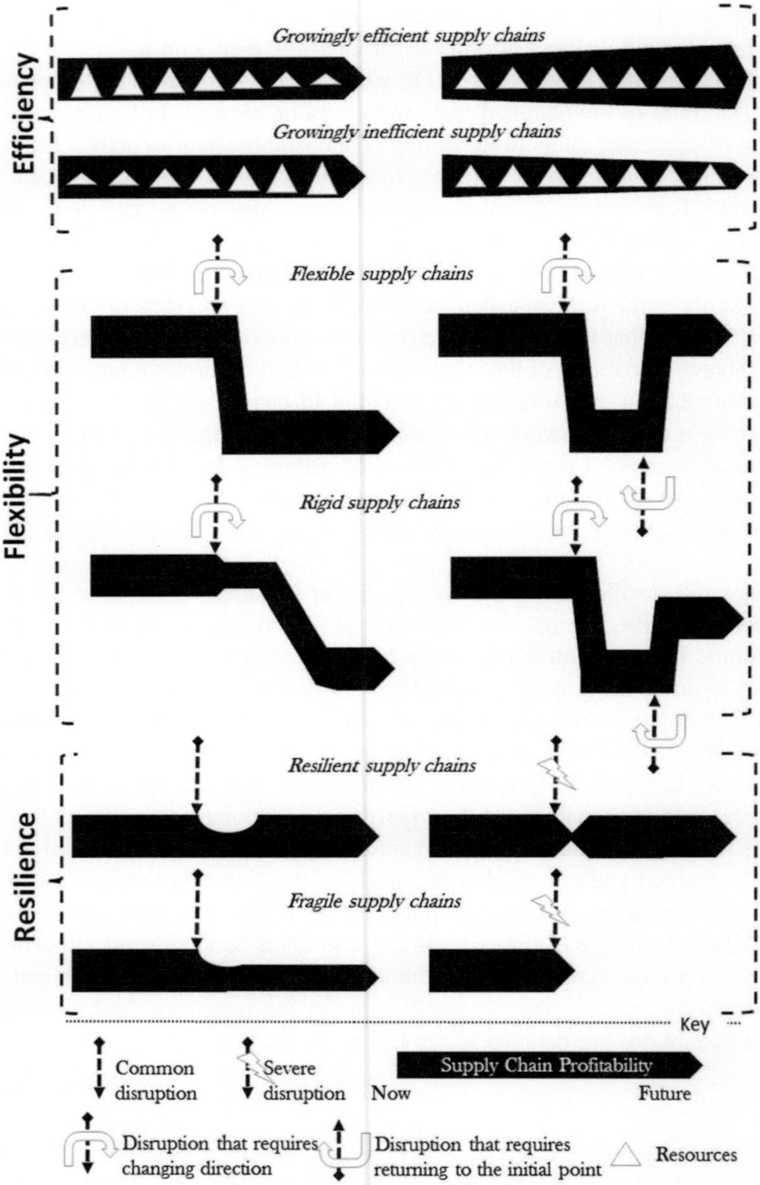

Fig. 2 The concepts of efficiency, flexibility and resilience applied to supply chains

2 Philosophical Approaches to Business Management: The Three Supply Chain Archetypes

The more and more demanding business context together with the core importance of SCM have encourage managers to look for strategies to improve the profitability of the supply chain by focusing on the three previously mentioned concepts: efficiency, flexibility, and resilience.

At this point, we must differentiate between two opposite approaches to SCM that can be observed in practice, representing two opposite philosophical strategies for problem solving: reductionism and holism (Ponte et al. 2016a). Reductionism involves breaking down the problem into smaller—and hence simpler—parts, which are analyzed and solved separately. Holism looks at the problem in its entirety, carefully taking into consideration the interdependences among its components.

In this sense, we understand the *traditional supply chain* (Holweg et al. 2005) as the paradigm of reductionism. In this production and distribution system, each node makes its own decisions aimed at maximizing local indicators based on its own information. Therefore, the global strategy of the system (that does not exist by itself) emerges as the interaction between the individual strategies of participant nodes that conform it.

On the contrary, the holistic approach leads to the *systemic supply chain*. Under this paradigm, the supply chain members participate in a collaborative process aimed at global optimization (Simatupang and Sridharan 2008)—it is therefore essential the use of overall, not individual, key performance indicators (1). This requires full transparency of the relevant information (2), as well as the complete integration of the processes throughout the system (3). The decision-making must be joint at the strategic level, while appropriately allocated in the planning and execution levels (4). In addition, the incentives must be aligned among the different supply chain nodes (5) in order to equally motivate all of them to care about the interest of the overall system.

In an intermediate point between both approaches, all kind of collaborative solutions that do not consider the five previously mentioned features, but only some of them, can be placed. For example, the *transparent supply chain* (Holweg et al. 2005), only incorporates information sharing into the system—commonly, point-of-sale data. A more advanced version is the *synchronized supply chain* (Ciancimino et al. 2012), which additionally merges the replenishment decision of each member with the production and material planning of its supply chain partners. Nonetheless, the various supply chain members keep a local perspective, focusing on their own metrics. Furthermore, the different supply chain actors are autonomous in terms of decision making; even if some operational decisions may be taken jointly.

Figure 3 displays the three main supply chain archetypes. The right part summarizes the basic information about them regarding the collaboration features. The left part represents each archetype using the simile of the gears, which represents the supply chain as a power transmission system between the natural raw materials and the final customer. Note that the two upper supply chains are driven by local

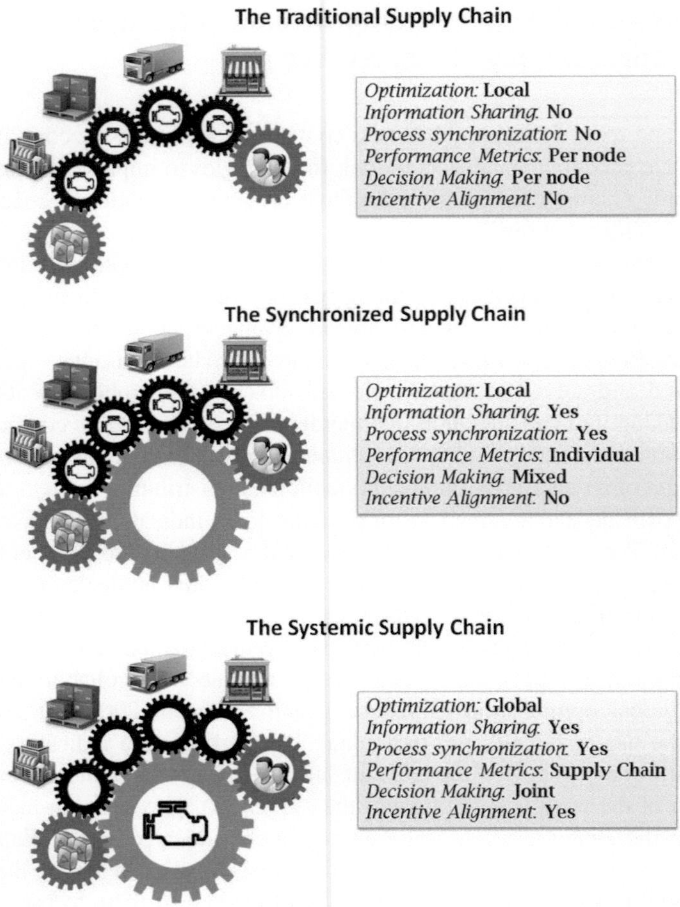

Fig. 3 The three supply chain archetypes using the simile of the gears

engines (one per echelon). The main difference is that, in the second one, there is an overall mechanism that keeps watch over and guarantees the synchronization of all the nodes. This mechanism also exists in the systemic supply chain. However, its role is prominent as it is the only one responsible for powering the system up.

3 Deploying the Systemic View: The Three-Legged Framework

Several studies have demonstrated how the traditional supply chain generates significant inefficiencies. For instance, Sterman (1989) analyzed the common results of the MIT Beer Distribution Game (Jairman 1963) and showed that the interaction of individual decisions produces aggregate dynamics which systematically diverge from the optimal behavior. This fact was illustrated by the amplification of the variability of orders and inventories as one moves up the supply chain as a consequence of the information distortion along the system. This phenomenon, the so-called Bullwhip Effect (Wang and Disney 2016), continues to be a major concern for real supply chain practitioners since it tends to increase different kind of costs (Ponte et al. 2016b).

From that point, the incorporation of areas of collaboration significantly improve the performance of the supply chain. For example, Hosoda and Disney (2006) showed that only by sharing information in the system the sum of storage and shortage costs can be reduced between the 8% and the 19%. Ciancimino et al. (2012) developed a model of a synchronized supply chain, which effectively deals with the Bullwhip Effect. The Vendor Managed Inventory (VMI) (Waller et al. 1999) and the Collaborative Planning, Forecasting, and Replenishment (CPFR) (Holmström et al. 2002) are examples of collaborative practices that have also shown to guide the supply chain toward significant operational—and consequently financial—improvements.

Nonetheless, the scope of the systemic supply chain is much broader. It implies the design of a totally collaborative supply chain aimed at optimizing the whole system. Precursors of this systemic orientation are Lean Management and the Theory of Constraints. Lean Management (Shah and Ward 2003) is the most well-known systemic philosophy. Focused on simultaneously improving the flow of value to customers and reducing the overall waste, Lean Management has completely revolutionized the production field since Toyota implemented it. Similarly, the Theory of Constraints (Goldratt 1990) is able to lead the production system to a dramatic improvement from a holistic perspective. Mabin and Balderstone (2003) estimated an average increase in the throughput by 63% without a significant grow in operational expense after reviewing real applications of this philosophy.

However, although these holistic methodologies are relative common from a single-company perspective, its application to the overall supply chain is far from being widespread. The conflict of interests between the different supply chain partners arises as a great barrier which obstructs the deployment of the systemic approach (Mentzer et al. 2001). That is, even though practitioners extensively recognize the supply chain as an overall system with growingly complex interdependences, they usually approach the issue from a local optimization perspective (Schneider 2013).

Under these circumstances, we highlight the systemic perspective as the great challenge of the 21st-century supply chains. Adopting a systemic view can lead the supply chain to major improvements—and consequently can strengthen the current global economic networks—, but the process of devising a *systemic supply chain* is tough, as it implies the consideration of multiple aspects in order to motivate all echelons towards the common goals.

This general challenge must be faced from three strategic pillars: education, orchestration, and methodology. This three-legged view is represented in Fig. 4.

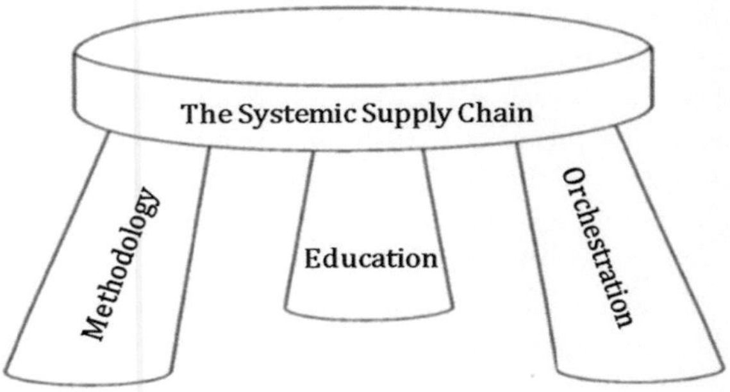

Fig. 4 The three-legged framework for deploying the systemic perspective in the supply chain

Education. Professor Sterman (1989, 2002, 2006), the Director of the MIT System Dynamics Group, has often highlighted that the underlying cause of the narrow deployment of the systemic perspective in the business scene is the lack of understanding of its counterintuitive concepts. This fact clearly points out to education as the foundation of the aforementioned challenge. A learning model based on analyzing the organizational issues in its entirety (focusing on the interdependences, the patterns, and the dynamics) must be favored in order to instill the need for 'thinking globally, acting locally' in the supply chain. The discussion of cases study together with the development of learning games, in an environment facilitated by the emerging technologies, can help lecturers to create this educational context. Prospective professionals must possess the required skills to deal with an increasingly complex and dynamic scene.

Orchestration. This pillar alludes to the design, development, and operationalization of the systemic solution in search of an efficient, flexible and resilient supply chain. At this point, we strongly recommend the conceptual framework by Simatupang and Sridharan (2005), based on five connecting features: (1) Information sharing, (2) Process integration; (3) Decision synchronization; (4) Overall scorecard; and (5) Incentive alignment. Puche et al. (2016) can be consulted to observe a collaborative scheme based on this framework and supported on the Theory of Constraints and the Viable System Model. This axe encompasses the coordination between theoretical foundations (like Lean Management and the Theory of Constraints), advanced technological tools (e.g. radio-frequency identification and real-time location systems), and collaborative practices (such as VMI and CPFR) under a robust collaborative scheme.

Methodology. Devising a collaborative solution requires a complex decision-making process, which must be supported through the appropriate techniques. In this sense, we strongly stand up for the use of modeling and simulation techniques; e.g. the emerging multi-agent systems (Gilbert 2008). These prototypes can be understood as powerful prototypes for business understanding and transformation as this risk-free environment would (Business Prototyping 2016): (1) Ensure that all members are involved in exploring new possibilities and designing concrete changes in the supply chain; (2) Help managers make the difficult questions and tentative answers, and hence support the development of the collaborative solution; and (3) Let practitioners see the big picture of the supply chain without losing sight of the details; and (4) Provide a blueprint for the evolution from a traditional to a collaborative scheme, leading to clear goals, concrete milestones and transparent progress.

4 Conclusions

Although managers widely recognize the supply chain as an overall system with complex interdependences, they do not usually operate consistently on that basis. In this sense, they usually approach the issue from a perspective based on local optimization that significantly damages the supply chain performance—for example, since these individualistic solutions fosters the Bullwhip phenomenon, which is a major concern for businesses. On the contrary, approaches built on global optimization have shown to lead the supply chain to breakthrough improvements. However, this kind of systemic approaches are far from being widespread, given that designing and implementing a successful collaborative framework in the supply chain is a convoluted process. Under this circumstances, we point out the deployment of the systemic approach as the great challenge for the 21st-century supply chains, which must be efficient, flexible, and resilient to deal with a growing variety of requisites.

In this work, we propose an approach far beyond mathematical optimization, where the intangible factors can be incorporated. We describe a three-legged framework for deploying the systemic (holistic) perspective in the supply chain. Education, orchestration, and methodology are its three strategic pillars. Education involves the development of a learning model based on analyzing the organizational issues as a whole (breaking down the 'divide and conquer' paradigm). Orchestration refers to the design and implementation of systemic solutions in the real-world, which considers both theoretical foundations, technological tools, and collaborative practices. Last but not least, a methodology to support the decision-making processes while adopting the collaborative scheme is essential to design a system with the desired properties and to align incentives between the different supply chain nodes. We acknowledge that this proposal has a conceptual value and practical experiences should take it into consideration.

References

Asbjørnslett, B. E. (2009). *Assessing the vulnerability of supply chains. Supply Chain Risk*. New York NY: Springer.

Beamon, B. M. (1999). Measuring supply chain performance. *International Journal of Operations & Production Management, 19*(3), 275–292.

Business Prototyping (2016) The benefits of business prototyping [online]. Available through http://www.business-prototyping.com/the-business-prototyping-manifesto/the-benefits-of-business-prototyping/. Last accessed April 28, 2016.

Ciancimino, E., Cannella, S., Bruccoleri, M., & Framinan, J. M. (2012). On the bullwhip avoidance phase: The synchronised supply chain. *European Journal of Operational Research, 221*(1), 49–63.
Fawcett, S. E., Ellram, L. M., & Ogden, J. A. (2014). *Supply chain management: From vision to implementation*. London: Pearson.
Gilbert, G. N. (2008). *Agent-based models*. London: Sage.
Goldratt, E. M. (1990). *Theory of constraints*. Croton-on-Hudson: North River.
Goranson, T. (1999). *The agile virtual enterprise*. Westport, CT: Quorum Books.
Holmström, J., Främling, K., Kaipia, R., & Saranen, J. (2002). Collaborative planning forecasting and replenishment: New solutions needed for mass collaboration. *Supply Chain Management: An International Journal, 7*(3), 136–145.
Holweg, M., Disney, S., Holmström, J., & Småros, J. (2005). Supply chain collaboration: Making sense of the strategy continuum. *European Management Journal, 23*(2), 170–181.
Hosoda, T., & Disney, S. M. (2006). On variance amplification in a three-echelon supply chain with minimum mean square error forecasting. *Omega, 34*(4), 344–358.
Husdal J (2008) Robustness, flexibility and resilience [online]. Available through http://www.husdal.com/2008/04/28/robustness-flexibility-and-resilience-in-the-supply-chain/. Last accessed April 27, 2016.
Husdal, J. (2010). *A conceptual framework for risk and vulnerability in virtual enterprise networks. Managing risk in virtual enterprise networks: implementing supply chain principles*. Hershey: IGI.
Jairman, W. E. (1963). *Problems in industrial dynamics*. Cambridge: MIT Press.
Mabin, V. J., & Balderstone, S. J. (2003). The performance of the theory of constraints methodology: Analysis and discussion of successful TOC applications. *International Journal of Operations & Production Management, 23*(6), 568–595.
Mentzer, J. T., DeWitt, W., Keebler, J. S., Min, S., Nix, N. W., Smith, C. D., et al. (2001). Defining supply chain management. *Journal of Business Logistics, 22*(2), 1–25.
Ponomarov, S. Y., & Holcomb, M. C. (2009). Understanding the concept of supply chain resilience. *International Journal of Logistics Management, 20*(1), 124–143.
Ponte B, Costas J, Puche J, de la Fuente D, Pino R (2016a) Holism versus reductionism in supply chain management: An economic analysis. *Decision Support Systems*, *86*(1), 83–94.
Ponte, B., Wang, X., de la Fuente, D., & Disney, S. M. (2016b). Exploring nonlinear supply chains: The dynamics of capacity constraint. *International Journal of Production Research,* in press.
Puche, J., Ponte, B., Costas, J., Pino, R., & de la Fuente, D. (2016). Systemic approach to supply chain management through the viable system model and the theory of constraints. *Production Planning and Control, 27,* 421–430.
Schneider D (2013). Systems thinking and the supply chain. Supply Chain Digest [online]. Available through, http://www.scdigest.com/assets/FirstThoughts/13-08-09.php?cid=7297. Last accessed April 26, 2016.
Shah, R., & Ward, P. T. (2003). Lean manufacturing: Context, practice bundles, and performance. *Journal of Operations Management, 21*(2), 129–149.
Simatupang, T. M., & Sridharan, R. (2005). An integrative framework for supply chain collaboration. *International Journal of Logistics Management, 16*(2), 257–274.
Simatupang, T. M., & Sridharan, R. (2008). Design for supply chain collaboration. *Business Process Management Journal, 14,* 401–418.
Sterman, J. D. (1989). Modeling managerial behaviour: Misperceptions of feedback in a dynamic decision making experiment. *Management Science, 35,* 321–339.
Sterman, J. D. (2002). All models are wrong: Reflections on becoming a systems scientist. *System Dynamics Review, 18*(4), 501–531.
Sterman, J. D. (2006). Learning from evidence in a complex world. *American Journal of Public Health, 96*(3), 505–514.

Tachizawa, E. M., & Thomsen, C. G. (2007). Drivers and sources of supply flexibility: An exploratory study. *International Journal of Operations & Production Management, 27*(10), 1115–1136.

Waller, M., Johnson, M. E., & Davis, T. (1999). Vendor-managed inventory in the retail supply chain. *Journal of Business Logistics, 20*(1), 183.

Wang, X., & Disney, S. M. (2016). The bullwhip effect: Progress, trends and directions. *European Journal of Operational Research, 250*(3), 691–701.

World Trade Organization (2016) International Trade Statistics 2015 [online]. Available through, https://www.wto.org/english/res_e/statis_e/its2015_e/its15_charts_e.htm. Last accessed April 27, 2016.

Chapter 7
Project Management Methodologies in the Fourth Technological Revolution

Javier Pajares, David Poza, Felix Villafañez and Adolfo López-Paredes

Abstract We are at the beginning of a new technological revolution, propelled by the development of cyber-physical systems and technologies like Internet of Things, Bid Data, Cloud Computing, 3D Printing, etc. Therefore, we will see an avalanche of projects to implement new business models, products, services and companies. In this paper, we analyse the main characteristics of these projects and we wonder about the appropriate methodologies and managerial styles to lead them. We argue that these projects are complex in nature, according to the current literature on project complexity and thus, classical project management approaches might be unsuitable for managing them. We suggest some clues to seek for new managerial styles, mainly in the literature concerning innovation and new product development and within the "Agile" approach.

Keywords Industrial revolution · Technological revolution · Internet of things · New technology projects · Project management methodologies

1 Introduction

We are in the edge of a new technological "revolution", probably a new technological step within the human being evolution.

The term "Fourth Industrial Revolution" was the main matter during the 46th Annual Meeting of the World Economic Forum (WEF, February 2016), where leaders from governments and private sectors met and discussed the challenges of the nearby future. Professor Klaus Schwab, founder of the WEF, outlined the technical and socioeconomic challenges the human being will face during the next decades. "*We are at the beginning of a revolution that is fundamentally changing the way we live, work, and relate to one another*" (Schwab 2016).

J. Pajares (✉) · D. Poza · F. Villafañez · A. López-Paredes
INSISOC. Social Systems Engineering Center, University of Valladolid, Pso del Cauce 59, 47011 Valladolid, Spain
e-mail: pajares@insisoc.org

C. Hernández (ed.), *Advances in Management Engineering*, Lecture Notes in Management and Industrial Engineering, DOI 10.1007/978-3-319-55889-9_7

Other words used by newspapers and consulting firms to name the new wave are "new digital economy", "interconnected world", "internet of things", "technological revolution", etc. The new technologies split into different areas of application, as cities (Smart Cities), factories (Industry 4.0., Smart Factory), services (Internet of Services), buildings (Smart Buildings) etc., affecting to almost every aspect of our lives.

The revolution consists in the development and massive utilization of cyber-physical systems, that is, the interconnection and communication of computing devices, physical objects, people and physical environments. The main idea is that many physical devices will be connected and communicate by means of internet. Common objects like freezers, boilers, traffic lights, clothes, etc. will be connected to internet and will communicate with central systems and within them, so that, these devices will become "intelligent".

The possibilities for new business models, new products and new services are infinite. This means that we expect to see an avalanche of projects: R&D projects for developing the new technologies, innovation projects for new products and services, projects for starting-up new companies, projects related to installation and tuning of equipment, maintenance projects, etc. The changes will affect to all sectors, giving rise to thousands projects in industry, buildings, homes, public sector, etc.

In this paper, we think about the methodological approaches for managing these projects. In particular, we wonder whether the traditional project management methodologies are appropriate, or alternatively, we should look for new methodological frameworks and managerial styles. To this aim, we review the classical project management hypotheses and its limitations. We focus on the literature that underlines that the classical approach fails whenever the project complexity increases.

Then, we highlight the main features of the projects of the Technological Revolution, focusing on the managerial implications and on the new skills that project managers will have to display, in a context of innovation and competitive pressure. We argue that, generally speaking, the traditional frameworks could not be appropriate to manage these projects, as they are basically "very complex" projects, in the sense described by the project complexity literature. In particular, to a greater or lesser extent, these projects exhibit high levels of structural and organizational complexity, high uncertainty, dynamic and sociopolitical complexity. Additionally, project management teams have to work under high pressure to finish the project as soon as possible.

We propose some clues to look for ideas to manage the new projects, mainly in the Agile philosophy and in the innovation and new product development literature.

The rest of the paper is organized as follows. First, in Sect. 2, we introduce the Fourth Technological Revolution and its applications in different sectors. In Sect. 3, we explain the main characteristics and main limitations of the classical approaches to project management, focusing on the role of project complexity. In Sect. 4, we describe some of the features of the projects of the technological revolution, mainly addressing the managerial implications. In Sect. 5, we argue why these projects are complex, as they exhibit many of the characteristics detailed in the project complexity literature. We propose to look for new methodologies and managerial styles in several areas, like Agile Project Management of the literature in innovation and new product development. Finally, we conclude with the main results.

2 The Fourth Industrial Revolution

In this paper, we mean "technological revolution" or "industrial revolution" as the integration of cyber-physical systems by means of communication between humans (C2C), humans and machines (C2M) and communication between machines (M2M) (Roblek et al. 2016). The IOT European Research Cluster defines Internet of Things as "*a dynamic global network infrastructure with self-configuring capabilities based on standard and interoperable communication protocols where physical and virtual things have identities, physical attributes, and virtual personalities, use intelligent interfaces, and are seamlessly integrated into the information network*". For the purposes of this paper, we use the interchangeably the terms "technological revolution", "internet of things" (IoT) or "new technology".

The revolution will take place because of the maturity of several technologies like big data, artificial intelligence, development and size reduction of sensors, robotics, new materials, 3D printing, cloud computing or biotechnology, among others. From a more technical view, Internet of Things relays on technologies like radio frequency identification (RFID), wireless sensor networks, middleware, cloud computing and application software (Lee and Lee 2015; Atzori et al. 2010; Whitmore et al. 2015, etc.)

Currently, we use internet for receiving and sending information (emails, knowledge, etc.), but the new digital era will allow to use internet for health purposes, shaving energy, managing traffic, etc. Until recently, only humans have introduced data in internet; now, also machines and common devices will send data to the global network.

The possibilities for new applications are infinite: Urban traffic systems that improve traffic by means of the communication between cars, traffic lights and parking lots; medical systems monitoring biological variables (blood pressure, sugar levels, heartbeat, etc.) of elder people living alone, so that an alarm signal is sent to the hospital in case of emergency; smart homes where heaters, lights, blinds, etc. are connected and "learn" the usual behavior of the residents, so that energy is saved, etc. In Table 1, we show some of the applications in different areas and sectors.

Both people and "smart devices" will produce a huge amount of data, which could be translated, by means of Bid Data technologies, into relevant information about customer behaviors, a valuable information for marketing purposes.

The terms Industry 4.0, Connected Industry or Smart Factory are commonly used to refer to application of the new technologies to the industrial sector. The introduction of the new technologies in the industrial sector will allow higher levels

Table 1 Some applications of internet of things

	Category	Example
Home	Smart Buildings	Security systems Energy efficiency Security
	Smart Home	Intelligent and connected appliances Home remote control and monitoring
Cities	Smart Cities	Public transport, parking, traffic management Energy saving, environmental monitoring Public safety
Industry	Industry 4.0	Production optimization, logistics Accurate predictive maintenance Improved retailing and supply chain management
	New products and services	Customization of products Product as a Service New after-sale services
Agriculture and Breeding	Agriculture	Weather monitoring Seed evolution monitoring Smart irrigation
	Smart Farm	Disease detecting and tracking Animal location Animal traceability
Business and Marketing	Marketing and new business models	Consumer behavior knowledge Consumer needs patterns Customized advertising
People lifestyle	Wearables	Smart watches Health monitoring and improvement Fitness and wellness products and services Elder people monitoring
	Self-productivity	New apps processing real remote data
	Tourism and leisure	Assisted driving, intelligent cars Augmented maps Improved social networks

of production optimization, better logistics (tracks, boats connected to factories and warehouses), and customization of products. Some products could be sold in the market as a service (Product as a Service), that is, the consumers will not pay for the "asset", but depending on how they use the asset. Broadly speaking, remote monitoring allows the concept of "anything as a service".

In a visionary paper published in the Harvard Business Review, Professor Michael Porter and James Heppelmann (Porter and Heppelmann 2014) explain how the new "connected products" will dramatically change the way firms work: from design to manufacture, from marketing to after-sale service. Companies will have to redefine their business, rethinking how to create value, how to get competitive advantages, who are the partners, etc.

It is said to be a "fourth" revolution to emphasize that its consequences will have the same magnitude than the three previous revolutions: first, the mechanical production thanks to the water steam; second, mass production powered by electricity; and third, industrial automatization by means of information technology. It is not a simple evolution of computing and information and communication technologies, but a revolution because of the social and economic consequences and the disruptive changes in value chains, production methods and business models. Business models will change because the new technologies will change key resources and activities, value propositions, costs and revenues structures and key partners (Kiel et al. 2016).

The figures concerning economic value of Internet of Things are impressive. McKinsey Global Institute (2015) forecasts that the potential economic impact will reach $11.1 Trillion per year in 2025. According to Price Waterhouse Coopers, around $6 Trillion will be spend on Internet of Things solutions between 2015 and 2020 (PwC 2016). The new technology will change how the existing companies work, transforming current business processes, and will inspire the creation of new products, services and business models.

3 Project Management Methodologies and Complexity

3.1 Classical Project Management

Although we can find projects since the ancient civilizations like Egyptians, Greeks and Romans, project management have not been developed as a discipline until the 50's, with the development of the CPM (Critical Path Method) and the PERT (Project Evaluation and Review Techniques) methods by the U.S. Department of Defense, within the context of the Polaris Program.

As a discipline, project management has been mainly developed by practitioners, who have gathered and structured "good practices", that is "*knowledge, skills, tools, and techniques* (which) *can enhance the chance of success over many projects*" (PMI 2013). Indeed, some of the most widely used bodies of knowledge have been developed by international professional associations, like the Project Management Institute (PMI) and the International Project Management Association (IPMA).

There is an historical divorce between the general management community and the project management community (practitioners and academics), probably because PM has been done by practitioners (mostly engineers) focused on the "task done culture", instead of the organizational cultures and skills.

The Project Management Institute (PMI) (pmi.org) was founded in 1969 in United States and currently it has around 500,000 members. PMI is organized by means of "Chapters" in cities around more than 200 countries. PMI publishes the

Project Management Body of Knowledge (PMBOK ® Guide), and it has issued the fifth edition in 2013 (PMI 2013).

The PMBOK develops a process-based framework of project management, that is, the guide describes the set of processes to be implemented in order to achieve project success. The fifth edition recognizes 47 processes that fall into five process groups (initiating, planning, executing, controlling and closing) and 10 knowledge areas (integration, scope, time, cost, quality, human resources, communication, risk, procurement and project stakeholder management). Process groups and knowledge areas are the two dimensions of a matrix structure, so that, each individual process belongs to a process group and a knowledge area.

The International Project Management Association (IPMA, www.ipma.ch) was founded in 1965 (the original name was INTERNET (INTERnational NETwork)) and it is made up of member associations, currently, more than sixty national project management associations around the world.

IPMA promotes a competence-based approach to project management, that is, defines the set of competences the project manager should have in order the project to succeed. According to IPMA, competence is "*the demonstrated ability to apply knowledge and/or skills, and, where relevant, demonstrated personal attributes*" (ICB 2006). IPMA competences are described in the International Competence Baseline (ICB). The ICB 3.0 considers three different groups of competences: technical, behavioral and contextual. Technical competences (20 elements) are skills and knowledge for initiating, managing the execution and closing a project. Behavioral competences (15) deal with the personal relationships between the individuals and groups managed in the projects; and Contextual Competences (11) refers to the interaction of the project team with the context of the project (other projects, permanent organization, etc.).

Recently, IPMA has issued the 4.0 version (IPMA 2015), which will substitute the previous 3.0 version. It considers 3 competence areas: People Competences (personal and interpersonal competences), Practice Competences (methods, tools and techniques used in projects) and Perspective Competences (interaction with the environment and project strategy and governance.).

The international standard ISO 21500 (International Organization for Standardization 2012) also adopts a process based approach, with 39 processes falling into five process groups (initiating, planning, implementing, controlling and closing) and 10 subject groups (integration, scope, resource, time, cost, risk, quality, procurement and communication) with similar roles than the PMBOK's knowledge areas. The Standard integrates ideas from different sources and project management bodies of knowledge, as PMBOK, ICB, PRINCE2, ISO 10006 or ISO 31000 (Stellingwerf and Zandhuis 2013).

Other popular bodies of knowledge, guides and methodological frameworks are PRINCE2 (Projects in Controlled Environments), developed and extensively used by the UK Government, P2M (Program and Project Management for Enterprise Innovation) proposed by the Project Management Association of Japan; APM Body of Knowledge by the British Association for Project Management (APM); etc.

Beyond the professional associations' frameworks, there is wide bibliography gathering the mainstream project management practices, methodologies and techniques.

Both professional bodies of knowledges and academic contributions on "mainstream" project management have adopted a systems analysis/systems management approach (Williams 2005), inspired on the seminal works by Cleland and King (1967, 1983). The best-seller book by Kerzner (2013, 11th edition, first edition in 1979), also emphasized the "systems approach" in its tittle.

3.2 Limitations of the Traditional Approaches: The Role of Complexity

Several authors have criticized the traditional approach to project management (see for instance, Shlin-Andersson and Söderholm 2002; Koskela and Howell 2002a; Morris 2004; Morris et al. 2011, etc.). This criticism is actually supported by the empirical evidence of generalized overruns in almost all projects in all sectors (see for instance, the CHAOS reports by Standish Group for IT projects (Standish Group, 2015) or Cicmil and Hodgson 2006), which can be translated into thousands of dollars lost by companies, governments and society.

The general idea underlying the criticism to traditional mainstream project management is that this approach cannot be effective when the project characteristics are far away from the hypothesis underlying the traditional bodies of knowledge and the systems approach to project management.

First, traditional project management gives a disproportionate emphasis to planning and control. It is supposed that both the current "state of the world" and the desired goals (the new state of the world) are absolutely known, so that, the plan can be easily deducted in order to lead the system from one state to the other. Moreover, planning is made easy by means of the decomposition of the problem into smaller parts, or chunks of work, usually independent -beyond sequential relationships- (Koskela and Howell 2002b). For this reason, "breakdown structures" (work, organizational, risk, etc.) are one of the most used tools for project planning and management. No matter how big the project is; just split up the problem into small parts, so that, each small piece of problem can be easily solved.

The emphasis on control derives from the emphasis on planning: checking whether the project is evolving according to plan, so that corrective actions could be taken whenever there are deviations from the initial plan. In case of big discrepancies, project managers seldom wonder whether the plan was wrong or if external conditions (like technology, stakeholder's needs, environment, etc.) have changed during project runtime.

Additionally, as far as we understand, human factors have not received enough attention within the classical frameworks. If planning and control are done properly, and all the information is available (at least in probability terms), the project team will lead the project to a successful end, as the persons are rational in nature and the goals and means are known. Of course, human resources issues are present in classical literature and anybody rejects the importance of leadership, motivation, communication, etc., but traditional approaches have not focused thoroughly on human "soft" skills. An exception is the competence approach by the International Project Management Association (IPMA), as it explicitly recognizes the group of "Behavioral Competences" (ICB 2006), or "Personal Competences" in the ICB 4.0 (ICB 2015).

Williams (2005) advocates that traditional project management methodologies can be inappropriate in projects characterized by complexity, uncertainty and tight time constraints (pace), and shows that current literature on project failure and success suggest similar parameters. He recommends moving to new project management methodologies where de project "emerges" rather than being fully preplanned.

Complexity is one of the critical aspects to take into account in order to determine the managerial approach to be used to lead the project to success (Baccarini 1996). Project complexity has received a wide interest by academics, as we can deduct by Geraldi et al. (2011). Not for nothing, Bosch-Rekveldt et al. (2011) report forty sources of complexity deducted from engineering project literature, etc., and a simple search in the International Journal of Project Management gave a result of 1205 papers in October 2016.

There are many definitions of complexity and complex project, but for the purposes of this paper, we follow de definition by Baccarini (1996), as "*consisting of many varied interrelated parts and can be operationalized in terms of differentiation and interdependency*". The idea of project complexity is aligned with the definition of complex system by Simon (1991): a system "*made up a large number of parts that interact in a nonsimple way. In such systems, the whole is more than the sum of the parts, not in an ultimate, metaphysical sense, but in the important pragmatic sense that, given the properties of the parts and the laws of their interaction, it is not a trivial matter to infer the properties of the whole*".

Williams (2005) alerts that traditional methodologies do not realize that complex and unpredictable project behaviors can take place during project runtime, because different parts of the project are not independent, and therefore, systemic phenomena appear because of the interrelations between the parts: casual feedbacks, non linear behavior, etc.

The idea of inter-relation among parts is crucial to project complexity. A big project can be "complicated" because of its large size, but it is not necessarily complex if, for instance, its scope is well defined and it can be decomposed in smaller independent parts. On the other side, a small project can be complex because of the conflictive interests among several stakeholders.

Williams (1999) proposes that project complexity has two dimensions: structural complexity and uncertainty. Structural complexity relates to the number (and heterogeneity) of different elements and the interdependence among these elements. Elements can be project deliverables, stakeholders, partner companies, work breakdown structure, tasks, etc.

Baccarini (1996) also distinguishes between organizational and technological structural complexity (Baccarini 1996): The former is related to the depth of hierarchy and heterogeneity of organizational structures, number of partner companies, heterogeneity of tasks, etc. Technological complexity depends on interactions between inputs, outputs, etc.

Uncertainty adds complexity to project management and is related to the degree of knowledge about the current and future state of the project. Turner and Cochrane (1993) distinguish between uncertainty in goals and uncertainty in project methods, and they suggest different project management styles depending on whether the goals and methods (to achieve the results) are well or ill-defined. Taking into account these variables, they define a 2 × 2 matrix with 4 kinds of projects. If goals are not well defined, it is necessary to negotiate them with the client, stakeholders, etc., whereas it is necessary to have brainstorming sessions for proposing the means to be used whenever they are not clear. When both goals and methods are bad defined, it is not possible to define properly the activities to be done, and therefore, the project must be addressed in terms of deliverables, milestones and control points.

Pace or tight time constraints is another factor for increasing complexity in projects. Tight time constraints involves more pressure on the project team. In addition, the necessity to reduce project times increases structural complexity. For instance, we can reduce time by overlapping several work packages, but in doing so, the interrelations between them will increase, and therefore, new systemic phenomena will emerge in the project (in the sense by Williams 2005).

Beyond structural complexity, uncertainty and pace, Geraldi et al. (2011) show how the project management literature has also included the dimensions of dynamic and sociopolitical complexity. Dynamic complexity relates to changes in the project, as changes in goals, context, technology, systems design, etc. The changes in the project and the resulting decisions by project managers can produce phenomena which is only observed far away in the future. Sociopolitical complexity emerges because of the interaction among people and organizations, like heterogeneous stakeholder's demands, people interests, etc.

We can summarize that classical Project management is adequate when there is not uncertainty in both project goals (scope perfectly defined) and means (well known procedures and methods); when there are no significant interactions among the different elements (technical, organizational, task, etc.) of the project (independently of its size); and there are no important changes in goals and means (for instance, technology, market needs, etc.) during project life cycle. Whenever these hypotheses do not held, traditional project management approaches could fail.

4 Featuring the Projects of the Technological Revolution

In this section, we describe some of the main characteristics of the projects of the Fourth Industrial Revolution. We focus on those features that could affect the management style when managing these kind of projects.

4.1 The Technological Map of the "Internet of Things"

Internet of Things (IoT) projects can involve many different technologies. Indeed, the technological revolution comes because of the maturity of several technologies and its empowerment by means of connectivity.

First, Internet of Things involves enabling technologies like identification and sensing technologies, communication, network technologies, network discovery, software and algorithms, hardware, data and signal processing, discovery and search engine, relations network management, power and energy storage, security and privacy and standardization (Bandyopadhyay and Sen 2011).

Currently, most of the projects adopt a Service Oriented Architecture with five layers (Atzori et al. 2010; Al-Fuqaha et al. 2015, etc.). These layers are: objects (sensors, devices), objects abstraction, service management, service composition, applications, and a transversal layer concerning the management of privacy and security (Atzori et al. 2010).

Porter and Heppelmann (2014) summarize the "new technology stack" that companies will have to develop in order to succeed in these projects. The stack includes new hardware and software embedded in the products (sensors, processors, operating systems, etc.), networks communication for allowing connectivity, and the product cloud (with databases, smart applications and algorithms) running in remote servers. Additionally, the stack contains transversal tools for integrating data from smart objects with the company business systems (ERP, CRM), tools for managing security issues and software for connecting with external sources of information.

Beyond the technologies above, the projects will also involve the complementary technologies that will allow create the final value to the product/service, like cloud computing, big data, artificial intelligence, robotics or 3D printing.

4.2 Pressure on Project Development Times

During the last decades, the product lifecycles have shortened dramatically. During the last century, a telephone model could be sold in the market during 10 or more years. However, one of the current smartphones becomes "obsolete" after 18 months. There is a strong competition between companies to issue the next model of tablet, phone, game console, a new cloud computing system, etc. The innovator not only wins a temporal monopoly, but a brand image and client fidelity over time. For this reason, sometimes it is more important to be the "first" in the market than to reduce costs in developing the new product or the new improved version of a product.

This phenomenon is especially important in technological products and it will be harder for the new technologies. Therefore, time will be one of the most important variables in project management, probably more important than cost and even than scope. If it is necessary, a newer version could be launched onto de market later, in order to fulfill new technical performance. Project managers and teams should be used to work under time pressure.

4.3 *Business Model Orientation*

The new technological revolution will drive new products and new creative business models. Usually, in traditional project management, the success of the project is related to the success in getting a "physical" deliverable, and usually in terms of the iron triangle (scope, budget and time).

However, in the new context, the success of the project is related to the success of a new product or service in the market. In particular, the concept of success should be understood in a multidimensional sense, as proposed by Shenhar et al. (2001, 1997), who consider dimensions like efficiency (the traditional view of success based on the iron triangle), benefits for the customer (added value), business success for the organization (profits, market share, etc.) and the capability of the project to prepare the company for the future challenges.

This implies that project managers should be focused on success in the market or in a value proposition, instead of being (physical) deliverable oriented. Moreover, in some projects, the objective will not be a specific deliverable, but in succeeding in a business model. In other words, the project manager will be more "business oriented" instead of deliverables oriented.

This means that sometimes, the project physical deliverable will not be know at the beginning of the project (uncertainty in goals), and therefore, "plan" and deliverables will reveal during project runtime depending on the result of project activities.

In the new IoT projects, the output of some activities is knowledge, new knowledge that should be used to take decisions about the new directions the project has to follow. The project describes a path, which is unfold during project runtime, and it is not predictable before the project starts. This means that some projects will need an iterative approach during runtime and, for projects in the border of current technology, it will be necessary to adopt a trial-and-error type management.

4.4 *Inter-relation Between Firms: Cooperating in a Network*

The technological complexity of the projects is very high, and requires a wide spread of heterogeneous technological expertise under the same project or the same program. For this reason, it is common that several companies will be working together in the same project, each one putting in its specific expertise and knowledge: for instance, net providers, hardware providers, software developers, experts in sensors, bid data analysisists, cloud computing, etc. A report by World Economic Forum (2016) emphasizes the role of multisector partnership, including public-private collaboration: Business will need to realize that "*collaboration on talent issues, rather than competition, is no longer a nice-to-have but rather a necessary strategy*".

Beyond technical expertise, many projects will need the engagement of persons and companies with multidisciplinary competences: lawyers concerned with legal issues related to data privacy, philosophers engaged in ethical issues, sociologists, psychologists, etc.

Outsourcing and working by projects will be a common issue in this kind of projects. However, while computer science or engineering companies are used to work by projects, other professions like lawyers of phycologist are not. Therefore, the heterogeneity of disciplines involved in the same project will increase the managerial complexity, and the project manager will need to develop cross-disciplinary management competences, particularly when companies involved in the project have different organizational cultures.

Concepts like innovation labs, business incubators, venture investments, etc. will be a comfortable environment for the new high technology projects. Most of the high technology projects will be developed in an "open innovation" environment, and therefore, project managers will have to develop competences in order to work within these contexts, where the manager will collaborate with other firms, persons, institutions, etc. that develop new innovative ideas outside the firm. The term "open innovation" was coined by Chesbrough (2006), recognizing that in a high technological world, firms cannot generate inside all the knowledge they need, and therefore, companies have to collaborate in order to combine internal and external ideas and paths to market; the new economic value is created with knowledge generated inside and outside the firm. Open innovation practices will be fully integrated in innovation practices (Huizingh 2011).

In other words, the project management of the technological revolution will work in new innovative ecosystems, different from the traditional in-company projects. Some traditional sectors as construction or software development are used to work leading several subcontractors or outsourcing firms. However, the new ecosystems will be featured by projects performed by a network of companies, without a central firm coordinating and leading the project. The concept of project portfolio will be redefined in terms of network portfolio, that is, a portfolio of projects deployed in order to implement the shared strategy of a network of companies, instead of a single firm.

We are moving from times where the project manager is the "owner" of the project during runtime, to new times where the project manager is the driving force of a net of collaborative firms, a person capable of generating trust among people and companies.

4.5 Managing Projects in Start-up Firms

The new opportunities will inspire the creation of new firms wanting to implement new ideas of new business models, products and services. Some of these firms will be spin-offs from existing companies, while others will appear from the energy of new (sometimes young) entrepreneurs.

A start-up is a new firm that is still looking for a validated business model (Blank and Dorf 2012). This means that there is a big uncertainty about the success of the new products and services offered by the firm.

Most of these technological companies adopt the lean start-up philosophy by Ries (2011) and the business model approach by Osterwalder and Pigneur (2010). In order to discover a successful business model, firms develop "minimum viable products (MVP)" to be tested in the market, so that the firms receive feedback form the market and adapt their business model with the new data. To this aim, firms create a MVP expending as less money as possible, and check what the potential users think about it. With this new information, they propose an improved MVP. Projects to develop a new product or service are put into practice by means of an iterative process, until a definite product is succeed in the market.

Most of the projects in the new technological revolution will be developed within this lean framework, and therefore, managerial styles should be aligned with this iterative approach to project management, where the final deliverable "emerges" from interactions with the market.

Pajares et al. (2016) show how to manage start-up technological firms as a portfolio of projects. The functional organization is totally replaced by a project-based organization. A person can be the project manager in one project (or program) and can work as a team member in other project. The portfolio approach allows firms to get flexibility and to develop a validated business model with minimum expenses.

5 Projects of the Technological Revolution Are Complex in Nature

Under the umbrella of technological revolution, we mean a wide number of different kinds of projects, with different purposes, size, budget, etc. For instance, we can find R&D projects, some of them in the edge of technology, projects for developing a new product or service, projects for installing a service or product, maintenance projects, etc. Obviously, these different types of projects require different managerial approaches, depending on variables like risk, novelty, size, etc.

However, in this section we want to conclude about general characteristics of the projects in the new technological revolution, wondering about the managerial approach to be used in these kinds of projects.

Table 2 The complexity of the new technology projects (in relation to project complexity literature)

Complexity dimensions		Project features and sources of complexity
Structural	Technological	Number and heterogeneity of technologies Multilayer architectures Sensors, cloud computing, big data, 3D printing
	Organizational	Multidisciplinary teams Multi-organization projects Open Innovation Innovative Ecosystems
Uncertainty	Goals	Business and innovation related project objectives Product characteristics bad defined
	Methods	High technology, sometimes in the edge of the "state of the art" Multidisciplinary projects: engineers, psychologists, lawyers, etc.
Pace		Tight delivery time because of competition Shorten product lifecycles
Dynamic		Changes in goals during project runtime Lean start-up approach. Iterative business model Iterative product/service Changes in technology during project runtime
Sociopolitical		Multidisciplinary views Legal issues: privacy, security Ethical issues Stakeholder heterogeneity

As far as we understand, complexity is the main characteristic of the new technological projects. In Table 2, we show the dimensions of complexity following Geraldi et al. (2011) and the literature described in Sect. 3.2. For each dimension, we specify the related project features and sources of complexity.

The number and heterogeneity of technologies involved in a IoT project contributes to increase structural complexity. Some of the technologies are close related to hardware developments while others are related to software, networks, etc. To some extent, organizational complexity is a consequence of technological structural complexity, as the heterogeneity of technologies involved in the project require the participation of different competences, and expertise from different companies. But additionally, non technical persons and firms like lawyers of philosophers, etc. will participate within the project to deal with ethical o privacy issues, which are a main concern in the new projects. In order to deal with the innovative and multidisciplinary nature of these projects, companies will do projects in new managerial environments, like open innovation or "projects in network". This is a new challenge for project managers, who have to move to different ecosystems.

A relevant degree of uncertainty (in the sense of Turner and Cochrane 1993) is usually present in these kinds of projects. First, the projects of the new industrial revolution are intensive in software engineering, and information technologies, which are usually characterized by bad definition in project goals and requirements (Kaur and Sengupta 2013; Hofmann and Lehner 2001; Verner et al. 2008; Sommerville and Sawyer 1997, etc.).

For projects in the edge of technology, both goals and methods are not completely known at the beginning of the project, and they emerge during project runtime, as the project "produce" new knowledge that is used to take decisions affecting the evolution of the rest of the project. In other cases, like new products and start-up companies, the final aim of the project is the success of a business model, and this objective can be reached by means of different "physical" goals, mainly discovered in an iterative way, within a lean innovation philosophy.

Furthermore, in this lean innovation context, the particular goals of the project can change, and even the technology also can evolve during project runtime, especially when the project uses pre-paradigmatic technologies. These issues increase the project dynamic complexity, in the sense by Geraldi et al. (2011).

Competitive pressure and product lifecycle reduction increases the pressure for reducing development times, and therefore, project time tightening is another source of complexity. Project management methodologies should adapt to these high competitive environments.

Finally, the projects in the new technological age will suffer from sociopolitical complexity. First because of the legal and ethical issues involved in the new projects. Internet of Things will generate a lot of data concerning how humans behave. Privacy will be a main concern. For instance, if home appliances are connected to the net, this information could be used by other people (including hackers) to know about lifestyles for marketing purposes, or even used by thieves to know whether the residents are at home. Other example: biological data gathered by wearables of thousands of people could be used to improved medical science and benefit everybody; but individual information could be used by insurance companies to ask for more money to persons with more "probability" to be ill.

Sarma and Girão (2009) underline the role of the "attackers" to compromise the interests of typical stakeholders like users, providers of services and infrastructure and the society. It is necessary to develop an international legal framework taking into account the right to information and the privacy needs of uses (Weber 2009, 2010).

While more dimensions of complexity are present in a project, classical methodologies in project management could stop being adequate for managing the project, and therefore, new frameworks should be used.

6 New Project Management Styles and Methodologies

The new technological revolution involves the creation of new products, new services and new business models. Therefore, these projects have a strong innovative component and we should look for the intersection of innovation management and project management in order to find keys to succeed. However, Shenhar and Dvir (2007a) highlight that although there is a lot of research concerning innovation management, there are very few studies on the linkage between innovation management and project management.

Furthermore, Keegan and Turner (2002) explain that project management is strongly concerned with project control, and project control can stifle innovation, and therefore, new product development. Elonen and Artto (2003) highlight some of the problems when managing innovation and development projects in multi-project environments. Some of the most common problems are lack of commitment, unclear roles and responsibilities, inadequate information management and bad project oriented culture.

In any case, the new technological projects will require an effort for a marriage between innovation and project management, if we want the projects to succeed, because "*successful innovation requires setting up well organized and well run projects that will put things in place in the most effective and efficient way*" (Shenhar and Dvir 2007a).

Pajares et al. (2016) show that a project management approach within a portfolio management environment can be a key factor of success and innovation in a technological start-up devoted to 3D printing technologies. Hobday (2000) shows the advantages of project based organizations for managing the development of complex high value products: customer orientation, flexibility, etc.

As we have argued above, the projects of the new revolution are basically complex. We described in Sect. 3.2 that classical methodologies cannot be adequate whenever the complexity associated to a project increases. This means that the project management community should search for methodologies that can deal properly with complexity. The question is now: Where do we look for these new frameworks?

The answer is not easy at all, because of the multidimensional features of the new technology projects described above in this paper. In a first attempt, we guess we could look at current alternative managerial frameworks for complex projects like the Agile framework and to the new literature concerning innovation and new product development.

Additionally, we suggest exploring the new professional competences to be displayed by the new technology project managers.

6.1 Complexity and "Agility"

In order to deal with complexity, there is an increasing literature encouraging project managers to move from a planning or instructional approach to a "learning approach" where the project "emerges" instead of being fully preplanned (Williams 2005), depending on the information obtained from the project during runtime and how the "unknowns" become "knowns" (Pich et al. 2002). This perspective has been implemented in practice by means of the "Agile" methodologies, widely used in software development.

The "Agile Manifesto" was written in 2001 for "*uncovering better ways of developing software by doing it and helping others to do it*". The main values underlying the manifesto are "*individuals and interactions over processes and tools, working software over comprehensive documentation, customer collaboration over contract negotiation and responding to change over following a plan*" (Manifesto 2001).

In fact, Agile comprises a set of different methodologies like Scrum, Extreme Programming, Feature Driven Development, Dynamic Systems Development, Crystal and Agile Modeling (Lindvall et al. 2002).

Within Agile philosophy, project team, customers and stakeholders work together iteratively, to discover the product requirements and functionalities. In each iteration, the team complete a quick planning–developing cycle, evaluating the evolving results by means of the feedback from the customer. This feedback allows to modify and improve the product and to discover new functionalities or goals during project runtime.

Thanks to this iterative procedure, Agile methodologies work well when the goal of the project is not clear (well defined) as it happens in most of the new technology projects. Under the Agile framework, both technology and requirements are allowed to emerge during the product development cycle (Lundavall et al. 2002).

Furthermore, as far as we understand, this philosophy is aligned with the lean start-up thinking described in Sect. 4.5, where the final product (or service) is obtained iteratively from the evolution of "minimum viable products" validated by the market. Each iteration adds value to the product until the final version is reached.

Although agile methodologies have advantages and disadvantages (see for instance, Serrador and Pinto 2015; Fernandez and Fernandez 2008) and some scalability problems (Boehm and Turner 2005), we think that they should be taken into account when developing new technology projects concerning internet of things, bid data, cloud, etc. Anyway, we encourage the project management community to explore new frameworks exploiting the learning-emerging approach, beyond the agile methodologies.

6.2 Technology-Dependent Project Management Styles

Obviously, not all the projects are equal and the management methods should be different depending on project size, complexity, technology, cultural issues, etc. Contingency Theory applied to innovation and new product development suggests that different internal and external conditions require different organizational settings to manage innovation projects. Sauser et al. (2009) report around 20 different frameworks for classifying and categorizing projects, proposing different managerial approaches to deal with them.

Among this literature, there are some classifications focusing on the role of technology, and therefore, they could be useful in order to find out methodologies and ideas to manage the projects in the new industrial revolution.

As far as we understand, Shenhar and Dvir (2007b) propose an interesting framework for dealing with most of the sources of complexity of the projects of the technological revolution. They suggest categorizing projects taking into account 4 dimensions: novelty, technology, complexity and pace. If we draw the dimensions into axes, a project is characterized by a shape reminding a diamond (see Fig. 1). The values in each dimension have important implications in the way we manage the project, especially novelty and technology in the case of the projects addressed in this paper; complexity and pace are common elements of this kind of projects.

Novelty is related to uncertainty in project goals and market, that is, we do not know ex-ante whether and how the new product (or service) will be accepted by customers, and therefore, the final characteristics of the product. Thus, high novelty usually implies that the project product and scope is not well defined, and an iterative project management approach is needed.

The technology dimension summarizes the technological features of the project. Low-Technology projects use only existing and well known technologies;

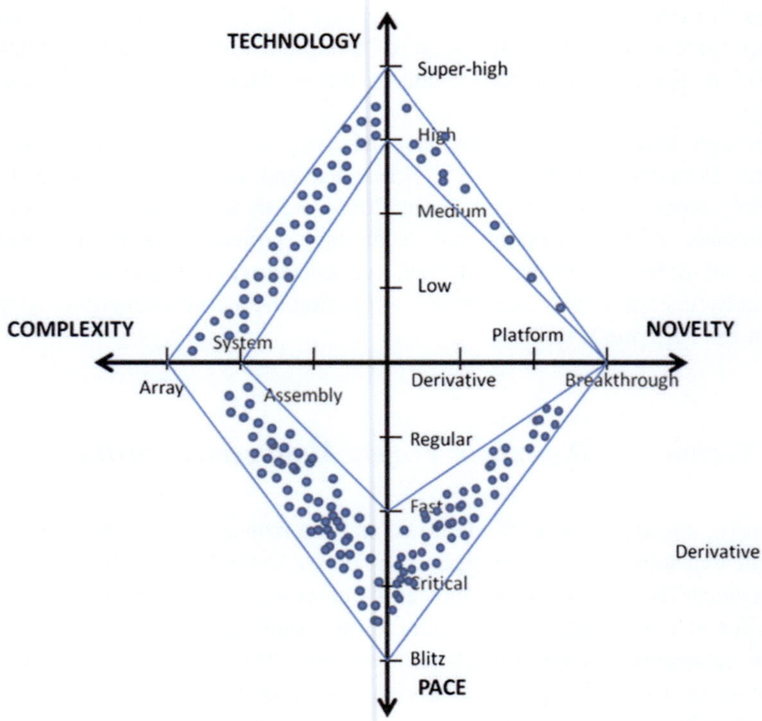

Fig. 1 The new technology revolution projects in the diamond framework by Shenhar and Dvir (2007a)

medium-tech uses mainly known technologies, but includes some technical characteristics not included in previous versions of the project; high-tec means that the project includes technologies already in the state of the art, but not previously used and known by the firm; and super-high-tec implies that some technologies will be developed during project runtime. The higher the technology dimension, the higher the technological skills required from the project team and the higher the technological uncertainty.

Complexity measures the complexity of tasks, products and organizations and pace relates to the urgency to finish the project. Complexity affects to the design of the project organization, whereas pace influences the required speed of decision and the autonomy of the teams in order to gain speed and flexibility.

In the diamond framework, the projects of the new revolution can be situated in the colored dots area in Fig. 1. The project will be usually characterized by high novelty, as the deliverable will be a new product or even a new business model. Technology is high and super-high as the new projects involve developing and implementing new technologies for the firm. Most of the projects imply the assembly of different systems (sensors, web, smart objects, etc.) and the competitive pressure will affect dramatically the urgency to deliver new products into de market.

Other source of ideas for managing the new technological projects comes from the innovation and new product development literature. In particular, there is and increasing literature in best practices for new product development (for instance, Cooper et al. 2004a, b, c; Kahn et al. 2006; Barczak and Kahn 2012, etc.).

Kahn et al. (2006) describe six dimensions of best practices in new product development: strategy, portfolio management, process, market research, people, and metrics and performance evaluation. Based on this research, Barczak and Kahn (2012) underlay some best practices by means of Delphi methodology with leading experts. As far as we understand, some of them can be easily translated into project management practices and strategies for managing the new projects. For instance, defining clear goals, implementing a project portfolio approach with good alignment between projects and strategy, participation of the customer/user during project runtime, top management support, good project leader and a stable project team.

6.3 Project Management Competences for the "Revolution"

Another relevant issue to address is the professional competences required for project managers in the new technological era. The competence approach is especially relevant in these projects, because they have almost all dimensions of complexity and uncertainty. The relation between project success and the personality and project manager's competences has been widely recognized in the literature (see for instance, Bakhsheshi and Nejad 2011; Crawford et al. 2005 or Crawford 2007).

As far as we understand, the project manager new competences will be related to complexity and to the new managerial styles they have to use in innovative environments.

Muller and Turner (2007, 2010) report the results of surveys answered by project managers, in order to find out the required leadership competences depending on project characteristics. Following the framework by Dulewicz and Higgs (2003, 2005), they consider intellectual, managerial and emotional competences. The results confirm general competences for all types of projects (no matter size, sector, complexity, etc.) like critical thinking (intellectual) and emotional competences like influence, motivation and conscientiousness.

More specifically for the purposes of this paper, they report that for very complex projects all the three types of competences reveal to be critical, whereas for medium complexity projects, the most important ones are critical thinking (intellectual); the management competences of managing resources, empowering, and developing; and the emotional competences of self-awareness, sensitivity, influence, and conscientiousness. This means that the higher the level of complexity, the higher level of competences required.

Beyond these "complexity focused" competences, we can deduct other set or competences and managerial skills from the arguments of this paper.

First, project managers will need to exhibit competences related to the characteristics of the people in the new project teams. Thus, the capacity to work with multidisciplinary teams is becoming more and more important, as the new projects involve different technologies and working with people with heterogeneous backgrounds as lawyers, psychologists, etc. In a global world, cross-cultural management will be crucial to manage projects involving people from different countries.

Second, the new innovative environments will require displaying new business oriented competences. The new technologies will represent some degree of disruption in business models. The project manager needs to be more business oriented than it is now, and he/she needs to move from focusing on the "iron triangle" to thinking in terms like economic value, strategic value, the urgency to issue a new product, implications of the project in firm's knowledge stock, etc. The connection between innovation and project management procedures will be essential.

Project managers will have to learn to work in distributed networks, either as a network actor or as a network leader. This way of working differs from being the leader of a net of sub-contractors and therefore represents a challenge for the project managers, who will have to work under the "open innovation" philosophy. Competences like trust and collaboration are essential.

7 Conclusions

The massive development and interconnection of cyber-physical systems is going to change the way people do business. The technological revolution will bring new products and services that will be developed thanks to technologies like Internet of

Things, Bid Data or Cloud Computing. These new projects will be run in new innovative ecosystems, with companies collaborating within a network and most of the times within new start-up firms that have not established a validated business model. In this environment, the pressure to reduce the time to introduce a new product or service into de market becomes crucial, and therefore, the need to improve project durations.

We have analyzed some of the main features of the projects in the new technological revolution and we have argued that traditional project management styles might be inappropriate whenever the complexity of the projects in high.

In this paper, we do not propose a new methodology for managing this project, but we suggest some clues to look for effective project management styles. In any case, the revolution of cyber-physical systems represents a new wave of innovation, and thus, project management methodologies should be aligned with some styles and practices from the fields of innovation management or new product development. We encourage the project management community to explore a convenient marriage between innovation and project management, in order to succeed.

In the new competitive environment, project managers have to exhibit new professional competences to deal with complexity and to work in decentralized environments, like open innovation or network collaboration. In other words, the project management community is going to face new exiting challenges, and new opportunities for improving strengthen and consolidate the profession.

References

Al-Fuqaha, A., Guizani, M., Mohammadi, M., Aledhari, M., & Ayyash, M. (2015). Internet of things: A survey on enabling technologies, protocols, and applications. *IEEE Communications Surveys & Tutorials, 17*(4), 2347–2376.

Atzori, L., Iera, A., & Morabito, G. (2010). The internet of things: A survey. *Computer Networks, 54*(15), 2787–2805.

Baccarini, D. (1996). The concept of project complexity—A review. *International Journal of Project Management, 14*(4), 201–204.

Bakhsheshi, H. A., & Nejad, R. S. (2011). 'Impact of Project Managers' personalities on project success in four types of project. *International Conference on Construction and Project Management, 15*(2), 181–186.

Bandyopadhyay, D., & Sen, J. (2011). Internet of things: Applications and challenges in technology and standardization. *Wireless Personal Communications*, *58*(1), 49–69.

Barczak, G., & Kahn, K. B. (2012). Identifying new product development best practice. *Business Horizons, 55*(3), 293–305.

Blank, S. G., & Dorf, B. (2012). *The startup owner's manual: The step-by-step guide for build-ing a great company*. Incorporated: K&S Ranch.

Boehm, B., & Turner, R. (2005). Management challenges to implementing agile processes in traditional development organizations. *IEEE Software, 22*(5), 30–39.

Bosch-Rekveldt, M., Jongkind, Y., Mooi, H., Bakker, H., & Verbraeck, A. (2011). Grasping project complexity in large engineering projects: The TOE (Technical, Organizational and Environmental) framework. *International Journal of Project Management, 29*(6), 728–739,

Chesbrough, H. W. (2006). Open innovation: The new imperative for creating and profiting from technology. Harvard Business Press.

Cicmil, S., & Hodgson, D. (2006). New possibilities for project management theory: A critical engagement. *Project Management Journal, 37*(3), 111.

Cleland, D. I., & King, W. R. (1967). *Systems analysis and project management*. NY: McGraw-HillNew York.

Cleland, D. I., & King, W. R. (1983). *Project management handbook*, New York: Van Nostrand.

Crawford, L., Hobbs, J. B., & Turner, J. R. (2005). *Project categorization systems: Aligning capability with strategy for better results*. Project Management Institute.

Crawford, L. (2007). Developing the project management competence of individuals. In L. Crawford (Ed.), *Gower handbook of project management*. UK: Gower, Aldershot.

Cooper, R. G., Edgett, S. J., & Kleinschmidt, E. J. (2004a). Benchmarking best NPD practices—I. *Research-Technology Management, 47*(1), 31–43.

Cooper, R. G., Edgett, S. J., & Kleinschmidt, E. J. (2004b). Benchmarking best NPD practices—II. *Research-Technology Management, 47*(3), 50–59.

Cooper, R. G., Edgett, S. J., & Kleinschmidt, E. J. (2004c). Benchmarking best NPD practices—III. *Research-Technology Management, 47*(6), 43–55.

Dulewicz, V., & Higgs, M.J. (2003). Design of a new instrument to assess leadership dimensions and styles. Henley Working Paper Series HWP 0311. Henley-on-Thames, UK: Henley Management College.

Dulewicz, V., & Higgs, M. J. (2005). Assessing leadership styles and organizational context. *Journal of Managerial Psychology, 20,* 105–123.

Elonen, S., & Artto, K. A. (2003). Problems in managing internal development projects in multi-project environments. *International Journal of Project Management, 21*(6), 395–402.

Fernandez, D. J., & Fernandez, J. D. (2008). Agile project management—Agilism versus traditional approaches. *Journal of Computer Information Systems, 49*(2), 10–17.

Geraldi, J., Maylor, H., & Williams, T. (2011). Now, let's make it really complex (complicated) A systematic review of the complexities of projects. *International Journal of Operations & Production Management, 31*(9), 966–990.

Hobday, M. (2000). The project-based organisation: An ideal form for managing complex products and systems? *Research Policy, 29*(7), 871–893.

Hofmann, H. F., & Lehner, F. (2001). Requirements engineering as a success factor in software projects. *IEEE Software, 18*(4), 58.

Huizingh, E. K. (2011). Open innovation: State of the art and future perspectives. *Technovation, 31*(1), 2–9.

International Project Management Association (IPMA) (2006). IPMA Competence Baseline, version 3.0. Nijkerk, the Netherlands: International Project Management Association.

International Organization for Standardization (2012). ISO 21500:2012. Guidance on project management. Geneva: ISO.

International Project Management Association (IPMA) (2015). The Individual Competence Baseline for Project, Program and Portfolio Management. International Project Management Association.

Kahn, K. B., Barczak, G., & Moss, R. (2006). Establishing a NPD best practices framework. *Journal of Product Innovation Management, 23*(2), 106–116.

Kaur, R., & Sengupta, J. (2013). Software process models and analysis on failure of software development projects. arXiv preprint arXiv:1306.1068

Keegan, A., & Turner, J. R. (2002). The management of innovation in project-based firms. *Long Range Planning*, *35*(4), 367–388.

Kerzner, H. R. (2013). *Project management: A systems approach to planning, scheduling, and controlling*, Wiley.

Kiel, D., Arnold, C., Collisi, M., & Voigt, K. I. (2016). The impact of the industrial internet of things on established business models. In *Proceedings of the 25th international association for management of technology (IAMOT) conference*, Orlando, FL.

Koskela, L. J., & Howell, G. (2002a). The underlying theory of project management is obsolete. In *Proceedings of the PMI Research Conference* (pp. 293–302). PMI.

Koskela, L., & Howell, G. (2002b). The theory of project management: Explanation to novel methods. In *Proceedings IGLC* (Vol. 10, pp. 1–11).
Lee, I., & Lee, K. (2015). The internet of things (IoT): Applications, investments, and challenges for enterprises. *Business Horizons, 58*(4), 431–440.
Lindvall, M., Basili, V., Boehm, B., Costa, P., Dangle, K., Shull, F., et al. (2002). Empirical findings in agile methods. In *Conference on extreme programming and agile methods* (pp. 197–207). Berlin Heidelberg: Springer.
Lundvall, B. Å., Johnson, B., Andersen, E. S., & Dalum, B. (2002). National systems of production, innovation and competence building. *Research Policy, 31*(2), 213–231.
Manifesto (2001). *Manifesto for agile software development.* http://www.agilemanifesto.org. 12 November 2016.
McKinsey Global Institute. (2015). *The internet of things: Mapping the value beyond the hype.* McKinsey Global Institute: McKinsey&Company.
Morris, P. (2004). *Current trends in project and programme management.* Association for Project Management Yearbook.
Morris, P. W. G., Pinto, J. K., & Söderlund, J. (2011). Introduction: Towards the third wave of project management. In P. W. G. Morris, J. K. Pinto, & J. Söderlund (Eds.), *The Oxford handbook of project management* (pp. 1–11). Oxford: Oxford University Press.
Müller, R., & Turner, J. R. (2007). Matching the project manager's leadership style to project type. *International Journal of Project Management, 25*(1), 21–32.
Müller, R., & Turner, R. (2010). Leadership competency profiles of successful project managers. *International Journal of Project Management, 28*(5), 437−448.
Osterwalder, A., & Pigneur, Y. (2010). Business model generation: A handbook for visionaries, game changers, and challengers. Wiley.
Pajares, J., López-Paredes, A., & Hernández, C. (2016). Technology start-up firms as a portfolio of projects: The case of DIMA 3D. *Procedia-Social and Behavioral Sciences, 226,* 59–66.
Pich, M. T., Loch, C. H., & Meyer, A. D. (2002). On uncertainty, ambiguity, and complexity in project management. *Management Science, 48*(8), 1008–1023.
PMI (2013). *A guide to the project management body of knowledge (PMBoK® Guide)* (5th ed.), Newtown Square, USA: Project Management Institute.
Porter, M. E., & Heppelmann, J. E. (2014). How smart, connected products are transforming competition. *Harvard Business Review, 92*(11), 64–88.
Price WaterHouse Coopers, PwC (2016). The industrial internet of things why it demands not only new technology—but also a new operational blueprint for your business. Price WaterHouse Coopers.
Ries, E. (2011). *The lean startup: How today's entrepreneurs use continuous innovation to cre-ate radically successful businesses.* Random House LLC.
Roblek, V., Meško, M., & Krapež, A. (2016). A complex view of industry 4.0. *SAGE Open, 6*(2), 2158244016653987.
Sahlin-Andersson, K., & Söderholm A., A. (2002). *Beyond project management—New perspectives on the temporary–permanent dilemma.* Copenhagen: Copenhagen Business School Press.
Sarma, A. C., & Girão, J. (2009). Identities in the future internet of things. *Wireless Personal Communications, 49*(3), 353–363.
Sauser, B. J., Reilly, R. R., & Shenhar, A. J. (2009). Why projects fail? How contingency theory can provide new insights—A comparative analysis of NASA's mars climate orbiter loss. *International Journal of Project Management, 27*(7), 665–679.
Schwab, K. (2016). *The fourth industrial revolution.* Geneva: World Economic Forum.
Serrador, P., & Pinto, J. K. (2015). Does Agile work?—A quantitative analysis of agile project success. *International Journal of Project Management, 33*(5), 1040–1051.
Shenhar, A. J., & Dvir, D. (2007a, August). The lost link: Why successful innovation needs sound project management. In *PICMET'07–2007 Portland International Conference on Management of Engineering & Technology* (pp. 597–610). IEEE.

Shenhar, A. J., & Dvir, D. (2007b). *Reinventing project management: The diamond approach to successful growth and innovation*. Harvard Business Review Press.
Shenhar, A. J., Dvir, D., Levy, O., & Maltz, A. C. (2001). Project success: A multidimensional strategic concept. *Long Range Planning, 34*(6), 699–725.
Shenhar, A. J., Levy, O., & Dvir, D. (1997). Mapping the dimensions of project success. *Project management journal, 28*(2), 5–13.
Simon, H. A. (1991). The architecture of complexity. In *Facets of systems science* (pp. 457–476). US: Springer.
Sommerville, I., & Sawyer, P. (1997). *Requirements engineering: A good practice guide*. New York, NY., USA: Wiley.
Standish Group (2015). Chaos Report. Standish Group.
Stellingwerf, R., & Zandhuis, A. (2013). ISO 21500 Guidance on project management-A Pocket Guide. Van Haren.
Turner, J. R., & Cochrane, R. A. (1993). Goals-and-methods matrix: Coping with projects with ill defined goals and/or methods of achieving them. *International Journal of Project Management, 11*(2), 93–102.
Verner, J., Sampson, J., & Cerpa, N. (2008). What factors lead to software project failure?. In *2008 Second International Conference on Research Challenges in Information Science* (pp. 71–80). IEEE.
Weber, R. H. (2010). Internet of things-new security and privacy challenges. *Computer Law & Security Review, 26*(1), 23–30.
Weber, R. H. (2009). Internet of things–Need for a new legal environment? *Computer law & security review, 25*(6), 522–527.
Whitmore, A., Agarwal, A., & Xu Da, L. (2015). The internet of things—A survey of topics and trends. *Information Systems Frontiers, 17*(2), 261–274.
Williams, T. (2005). Assessing and moving on from the dominant project management discourse in the light of project overruns. *IEEE Transactions on Engineering Management, 52*(4), 497–508.
Williams, T. M. (1999). The need for new paradigms for complex projects. *International Journal of Project Management, 17*(5), 269–273.
World Economic Forum (2016). The Future of Jobs: Employment, Skills and Workforce Strategy for the Fourth Industrial Revolution. January 2016. World Economic Forum.

Chapter 8
Knowledge and Innovation in the New Industrial Organization

Lourdes Sáiz-Bárcena, Miguel Ángel Manzanedo del Campo and Ricardo del Olmo Martínez

Abstract The objective of this work is to study "The New Industrial Organization" (Hernández 1997) and its implications for knowledge, cooperative relations and business innovation. This objective implies understanding the close and meaningful connection between Economics, Business Organization and Strategic Management. In doing so, we point out that two intangible factors, knowledge and cooperative relations between the agents of the firm are the basis of business profitability, through their impact on innovation. Innovation exists to convert knowledge into products and services that will attract the customers. Competitiveness emerges here, when those goods and services are unique and generate value for the customers, who are willing not only to acquire them, but also to pay the set price, and ultimate being an agent of the firm. The challenge consists in offering, today as ever, products and services that are yet to be found on the market, so that the firm will not be marginalized and abandoned by the customers, now a an agent of the firm's network. Knowledge, together with the innovative capabilities of people, opens up an enormous space for the firm, which leads to continuous progress full of opportunities and future.

Keywords Knowledge · Innovation · New Industrial Organization

---

L. Sáiz-Bárcena (✉)
Departmento de Ingeniería Civil, Escuela Politécnica Superior,
University of Burgos, Avda. Cantabria, S/N, 09006 Burgos, Spain
e-mail: lsaiz@ubu.es

M.Á. Manzanedo del Campo (✉) · R. del Olmo Martínez (✉)
Departmento de Ingeniería Civil, Escuela Politécnica Superior, University of Burgos,
C/Villadiego, S/N, 09001 Burgos, Spain
e-mail: mmanz@ubu.es

R. del Olmo Martínez
e-mail: rdelolmo@ubu.es

C. Hernández (ed.), *Advances in Management Engineering*, Lecture Notes
in Management and Industrial Engineering, DOI 10.1007/978-3-319-55889-9_8

1 Introduction

What does your firm produce or serve? This was a common question when two CEOs met at a conference in the eighties. Clusters of firms of the same sector gathered to share experiences and the portfolio of cases. Accordingly, Industrial Organization was concerned with sectorial economics, industrial dynamics, and with how to align the firm's strategy with the sector's opportunities. They met again in the nineties and the question probably was: what are your skills and competences so that we could cooperate? Industrial Organization had to deal also with governance and organizations structure: the skills, competences and cooperative sharing of non-tangible factors such as knowledge and joint innovations. If they met again in the last decade after congratulate for their good health, the professional question probably was: with whom do you network and have joint projects? A New Industrial Organization was emerging to cope with these new issues.

As C. Hernández noted, a series of influential forces have been redefining the firm and the way of understanding economic activity, under the name of the "New Industrial Organization" (Hernández 1997). These forces for change include the structure of the industry (structure-behavior-results), the institutional environment and the business governance structure and joint projects. The development of each force is linked to a vector that may, respectively, be termed: (1) the Vector of Market Opportunities and Market Threats; (2) the Institutional Vector; and, (3) the Governance and Operations Vector. Each of these contains a set of distinctive elements and factors, which, once developed, complete the objective and lead to a New Industrial Organization. One of the components in the vector of Governance and Operations of the firm is the "accumulation of knowledge and capabilities".

With this frame in mind, competitive advantage that implies the accumulation of knowledge and capabilities for the firm is analyzed. Not only accumulation, but also the need for the firm to gather, generate, share, transfer and permanently update the knowledge needed to meet its mission and strategies, to progress and to move ahead of its competitors (Sáiz-Bárcena et al. 2014).

An understanding of the firm as a collaborative activity between the agents that intervene endowed, above all, with knowledge and experience, leads to valuing the most intellectual dimension of all workers, as well as the resources of intangible nature; changing, in this way, the traditional conception of the firm. Now the firm is a set of high-level knowledge and closely-knit ties between its participants and stakeholders (Janhonen and Johanson 2011).

This new perspective requires strategies, organizational structures and cultural values that manage to turn it into a reality. As well as workers, it covers the design of tasks and objectives to put knowledge and profits, perhaps better described as benefits, generated by cooperative relations into action. To do so, union and perfect harmony between the sciences of economics, business organization and strategic management are necessary (Rumelt et al. 1991). Appropriate use of knowledge is the principal source of differentiation in an increasingly competitive and global market (Grant 1991; Schoemaker 1992). The main competitive advantage comes from the creation, acquisition, storage and diffusion of knowledge (Nahapiet and Ghoshal 1998).

Such advantages are the ones that frequently cause a quite broad difference between the quoted value of the firm on the market and its book value. This difference includes, precisely, the resources or components of an intangible nature and, in particular, the knowledge that allows the firm to develop competitive advantages in a permanent and sustainable way over time.

The structure of this work contains three related parts, which, all together, prompt reflection and bring us close to some of the aspects that can contribute to gaining a little more familiarity with the New Industrial Organization and its development in the reality of the today's firms.

The first section approaches knowledge and cooperative relations that the firm should deploy in the context of the new guidance provided by the perspective of the New Industrial Organization. Note that knowledge and cooperative relations between the agents of the firm are the basis of business profitability, arising from the generation of externalities within the firm, growing returns and avoiding the decreasing returns of traditional assets.

Following a logical sequence, the second section of this work is the move to the process that leads from knowledge to innovation. Innovation is the "art" of converting knowledge into products and services that fascinate and links the customers to the firm. Competitiveness emerges here, when those goods and services are unique and generate value for the customers, who are willing not only to acquire them, but to pay the set price. This value should be greater than the assets of other competitor firms and the challenge to do so is to offer, today and at all times, products and services that are still non available so that the firm is not seen to be abandoned by the market. To this end, the customer should be part of the firm's "collective intelligence".

In the third section, we discuss the implications of the New Industrial Organization frame for knowledge and innovation. This novel approach allows understanding better the behavior of the firm that will lead to a full exploitation of knowledge, cooperative relations and the capability to generate unique and competitive products and services. A close link should be deployed between Economics, Business Organization and Strategic Management.

2 Competitive Knowledge in the New Industrial Organization

The qualification and experience of workers, their capability to share and to update knowledge (Sáiz-Bárcena et al. 2013), and the web of business ties in place, to know at any point in time who the experts are and the necessary pieces of knowledge, are the fundamentals which sustain the New Industrial Organization.

In this way, human resources policies, defined to exploit the immense potential that people have, together with their knowledge, capabilities and behavior (Buller and McEvoy 2012), becomes an essential element for the selection, retribution and training of more qualified workers (McIver et al. 2012). This approach is one of the best ways of creating and maintaining distinctive competitive advantages; and to stand out in value creation in the eyes of present and future customers of the firm.

The process of sharing and updating knowledge is not free from difficulties, even in firms committed to this resource. Challenging problems of knowledge exchange and its obstacles arise, when the optimal assignation of resources or task completion requires specific knowledge localized in other individuals and groups.

Another important challenge is that people in general has no innate aptitude for knowledge sharing. When this knowledge is profound, sophisticated and complex, it is even more difficult. There are situations in which even the expert deliberately hides the necessary knowledge and denies its existence (Connely et al. 2012).

Workers may perceive a possible loss of recognition, in the form of promotions, increasing earnings or other fringe benefits, to which they consider they hold preferential rights, because they possess that unique knowledge. This situation may negatively affect their performance, as well as the relevance and uniqueness of such knowledge that loses the very attributes that had made it valid (Renzl 2008).

The design of the organizational structure, the decision-making framework and the managerial commitment of the firm can largely favor the effectiveness and the success of the relations, to ensure the availability of the necessary experts and knowledge (Mládková 2012). These factors also serve to root values such as confidence, reciprocity and mutual cooperation, the recognition of new ideas and solutions, the identification of high-value knowledge, its diffusion as well as the experience of feeling at ease with the processes in which valuable knowledge is put to good use and shared.

Unlike physical resources, which are consumed with use and imply decreasing performance over time, a notable attribute of knowledge is that it is one of the few resources of the firm that increases with use; an externality within the firm. The more knowledge is used, the more valuable it is and the more competitive advantage it can offer. And, although its creation is costly, its diffusion is easy thanks to the extensive development of Information and Communications Technologies (ICT). Therefore, knowledge intensive products and services show increasing yields; in other words, once the first unit is produced at significant cost, additional units may be generated at very low marginal costs.

Heightened performance means that firms grounded in knowledge and cooperation tend to be potentially more competitive, in great part, explaining a higher market value of the firm. Even more if knowledge is shared trough the "internet of things" by appropriate communication networks, a zero marginal cost goal will not be just science fiction (Gershenfeld 2011) and (Rifkin 2014).

3 From Knowledge to Business Innovation

Competitiveness is the capacity of the firm to maintain its advantage over its competitors for as long as possible. This competitiveness, together with other business elements, impacts in a direct way on profitability and affects products and services that should be more attractive for customers than those offered by other business competitors.

The question is, therefore, how to produce those goods and services that fascinate the market, which amounts to the transformation of knowledge into value. It is here that innovation enters into play, which undoubtedly requires knowledge. But that alone is not sufficient; it is its transformation into something that gives value to the customers. This is not an easy path; nevertheless the firm and its workers should travel down the road of innovation; as Einstein said, "Insanity: doing the same thing over and over again and expecting different results."

Innovation implies generating ideas that were not held before, which is why the key for innovation lies in asking the right questions, more than in looking for answers. In this sense, organizational culture can in a powerful way contribute to the development of innovative capabilities of the firm, because although we were born as "inquisitive beings", we have been educated to find answers. We have once again to recover the "art of the question".

One of the most relevant elements of intelligence is its learning capability, but to learn you have to think and the questions and doubts that represent the beginning of innovation arise, precisely, through the act of thinking intensely. It is difficult to innovate because, in general, the lives of many people, including their education and, if applicable, their work experience, revolve around what is already known—they are exposed at school and university to answers to what is known and discovered by others, while the proposal should on the contrary be questions without answers, the answers to which a priori are not known.

The challenge of firms, today and in the future, is not to provide products or services that already exist; in other words, "mistaken" products or services, but to imagine goods that are not in the market. To do so, firms should radically change their human resources policies to select or to develop highly curious, imaginative workers, with the desire to understand the environment and a commitment to permanent learning. Without doubt, learning is based on proposing appropriate questions.

Innovation, as understood in the New Industrial Organization, is a learning and path dependent process that begins when the firm decides to take an unknown route, in the form of new products and services. A path that gradually shakes off the uncertainty that is attached to it as the more appropriate questions are released, based on different aspects, such as the curiosity of the individual, observed anomalies, inconsistencies or challenges to what already exists. These are questions that emerge from the imagination, confusion or conflict and, most importantly, that require analytical capability, a critical spirit and experimentation, among others.

Evolutionary arguments are an appropriate approach to the analysis of industry dynamics in a knowledge-based economy, because they can deal properly with innovation processes, technological change, path-dependence and knowledge. Nelson and Winter (1982) and many others since them, advocated the use of computer simulation in order to design, build and run dynamic evolutionary models. Pajares et al. (2004) presented an agent based model to reproduce by simulation the main accepted facts of industry dynamics. Nevertheless to address a wider audience we keep the usual verbal accounting to reason about the role of knowledge and innovation in the N.I.O as part of Management Engineering.

A further question to highlight in innovation activities is that the questions that the worker formulates, necessary to scale these business heights, should demonstrate both the knowledge they hold and the knowledge they do not hold. In addition, those formulating the questions play the leading role and the initiative, which amounts to moving ahead to new products and services of high value. And more than that, it provokes, in turn, new questions and the cycle is repeated in a continuous virtuous circle, permanently generating the possibility of new assets.

In short, the innovative capability of workers opens a sustainable progress of the firm by sharing the collective intelligence of the firm and its customers, so that the market will not distance itself and leave the firm behind.

4 Implications of the New Industrial Organization for Knowledge and Innovation

Productive factors will not explain all economic growth. Solow (1956) attributed the unexplained part of this wealth creation to technological change (including education, research, innovations, etc.). This welcome residual implies that the aggregated productivity curve is moving upwards in such a way that in the long run, the area of decreasing returns is never reached, within the life cycle span.

Since then, many works have sought to explain the causes of that difference, known as the Solow's residual. In macroeconomists, the issue was settled by introducing in the aggregated production function leverage: the Total Factor Productivity (TFP). But in Management Engineering let it at the firm level or Industrial Dynamics the task was to identify the components of this TFP; the drivers of endogenous growth. In this context, Romer (1986) and Lucas (1988) proposed two ways of explaining the Solow's residual. The former stated that the capital (including the human capital) has two effects; a direct effect on the capital stock and an indirect one on knowledge accumulation. The latter considered the economy as a producer of tangible goods using, on the one hand, capital services and work (greater endowment of productive factors) and, on the other, intangible goods, and knowledge production, using investments in R&D.

In both studies, the conclusions were the same: the Solow's residual is fundamentally due to the capability to improve the intangibles of the firm and, in particular, to the accumulation of knowledge (Lucas 1993). The conclusion coincides with the view of the firm as a set of resources. As professor C. Hernández (op. cit) said, "the managerial and business capability to mobilize tangible and intangible resources would, in the last place, be the complete explanation of the Solow's residual. The relation between technology and growth is complete".

In standard management terms, firms seek sustainable competitive advantages, through strategies based on costs or differentiation. The creation of knowledge is decisive for both. In one case, the improvement of processes and learning by accumulation imply a reduction of costs. On the other, product differentiation increases the market share and adds value with respect to competitors, thanks to the income coming from the quasi-monopoly that it creates.

What are the firm's drivers that generate these two advantages? There are two levels of analysis of the firm's competitiveness. Both are relevant to answer the question. The traditional one from Managerial Economics and Industrial Organization, which analyzes the relations between the structure of the sector and the results for different options of the market structure, the strategic groups, entry barriers and mobility, the degree of rivalry and agreements, the negotiating power of suppliers and customers, etc. The opportunities focus. We are dealing here with the other one: the internal factors, resources capabilities and competences of the firm.

This approach allows us to determine the competitiveness of the sector more than of the firm and will not allow us to know why some firms are competitive and others less so, with the same technology, localization and other determinants that characterize the market. "The *heterogeneity* of the firm is, in the final analysis, the explanation of its sustainable competitive advantages. In this sense, the Industrial Organization approach helps us, in the best scenario, to diagnose sectoral competitiveness, but it does not tell us too much about how to improve it and, of course, almost nothing about how to manage research successfully to achieve increased competitiveness of the firm" (Hernández and del Olmo 1994).

The conclusion is simple, as Rumelt et al. (1991) pointed out, the connection between Economics, Organization and Strategic Management of the firm up until now is not fully developed, because economists resist the "micro-micro" study of the business unit and usually focused to sectoral studies. In any case, we cannot forget that the Microeconomics is neither a theory of the firm nor much less so a theory of firm management. Business strategy is neither an applied Microeconomics nor Industrial Economics; they are a basic component of the N.I.O, but the strategy contains hypotheses on organizational behavior, technology (development and diffusion of innovations), the 'values', the visions and the criteria for the future on the basis of a broad variety of knowledge. When "beliefs" and aptitudes are important, other disciplines are necessary.

A second approach, (Dosi 1988; Pavit 1990; Buesa and Molero 1993) characterizes the process of technological innovation; from research (knowledge creation) to concrete improvements in the profitability of the firm. This is a 'top-down' view, in which the firm is organized as a set of resources and capabilities, the successful mobilization of which is evidence of management capability.

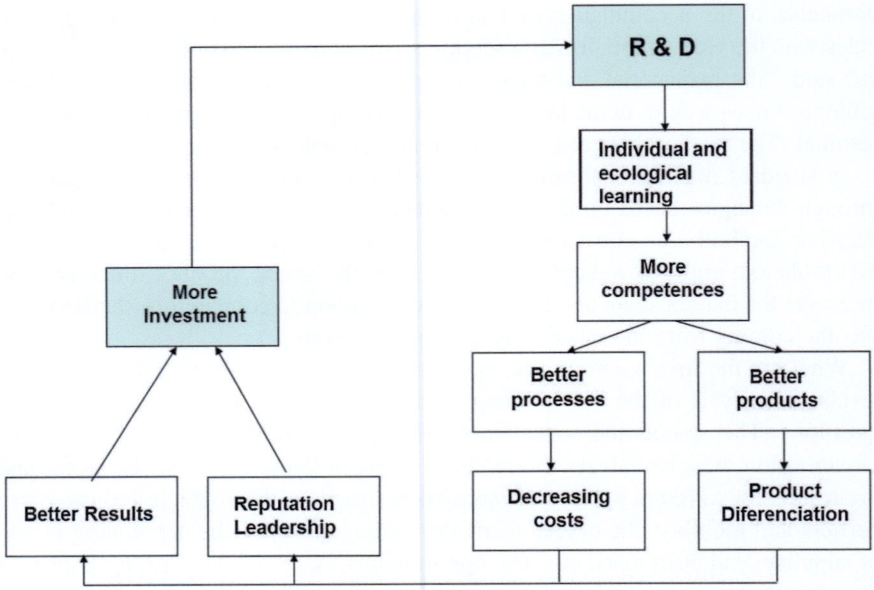

Fig. 1 R&D and business competitiveness

When adopting this approach, some characteristics of this process emerge that are implicit in the virtuous cycle of R&D and competitiveness (Fig. 1), (Hernández and del Olmo 1994).

As Porter (1990) states, competitive discipline and the business environment, in particular suppliers and customers, influence innovation capability. Knowledge creation, through R&D, affects the whole value chain, in particular the activities with greater value added potential.

The consequence is that the competitiveness of the firm requires an accurate categorization of firm activities, externalizing some of them if necessary; to create business networks that integrate suppliers and distributors and developing the industrial fabric of which the firm forms part.

The innovation process is very complex and investment in research is cumulative. Knowledge, whether scientific or technical, is not totally appropriated without costs. It is a codified knowledge, tacit, and specific to each firm and requires personal exchanges for its transmission.

Investments in intangibles, among which we may highlight organizational methods, are fundamental for the success of the investment in research. The innovation diffusion process is more than imitation. Both exploitation and exploration are needed.

The innovation process is interactive. Technological capital and human capital present synergies and, therefore, learning is as important as the incorporation of technology.

The culture of innovation and intangible value is something intrinsic to the culture of the firm. So, production decisions, JIT, MRP, CIM, etc., simultaneously affect the whole business organization.

This approach to innovation therefore highlights the conception of the firm as a set of resources. The success of the innovation process not only depends on what we call knowledge creation (on the basis of R&D), but on the other resources of the firm; the training of human resources, business organization and that visible hand called "entrepreneurship" are of special importance.

5 Conclusions

Management Engineering needs a new focus of Industrial Organization. In this study, knowledge, cooperative relations and the capability of business innovation have been analyzed taking as a reference the implications for the firm of the "New Industrial Organization".

This novel approach to Industrial Organization that has scarcely been studied requires us to know, to understand and to put into practice the union of Economics, Business Organization and Strategic Management taking advantages of the communications revolution and the digitalization of both, things and knowledge.

Thus, it may be concluded that a New Industrial Organization is needed to support Management Engineering. That knowledge, cooperative relations and business networks represent deciding factors in the profitability of the firm, as they generate better performance of great importance for its survival and sustainability. Cooperation and competitiveness are not rivals.

Innovation is the deciding factor in competitiveness, as the former manages to transform knowledge into products and services demanded by the customer, while the latter attaches value to those assets, so that the market selects those and not others.

The challenge for the firm is to be capable of offering new products and services, in a permanent way. These goals can only be achieved by growing knowledge, innovation and cooperation. It is here where "The New Industrial Organization" should deploy all of its potential, leading to greater profitability, the increase of competitiveness and the advance of business sustainability.

References

Buesa, M., & Molero, J. (1993). Tipologías de las estrategias innovadoras. *Economía Industrial, 289,* 49–62.

Buller, P. F., & McEvoy, G. M. (2012). Strategy, human resource management and performance: Sharpening line of sight. *Human Resources Management Review, 22,* 43–56.

Connely, C. E., et al. (2012). Knowledge hiding in organizations. *Journal of Organizational Behavior, 33,* 64–88.

Dosi, G. (1988). Sources, procedures, and microeconomic effects of innovation. *Journal of Economic Literature, XXVI*(3), 1120–1171.

Gershenfeld, N. (2011). *The coming revolution on your desktop-from personal computers to personal fabrication*. www.Readhowyouwant.com

Grant, R. M. (1991). The resource-based theory of competitive advantage: implications for strategy formulation. *California Management Review, 33*(3), 114–135.

Hernández, C. (1997). Crecimiento económico, progreso institucional y factor "e". España en la Unión Monetaria. Lección de Apertura del Curso Académico 1997–1998. UVA, Valladolid.

Hernández, C., & del Olmo, R. (1994). Investigación y desarrollo en Castilla y León: Análisis y orientaciones estratégicas. In A. Hernández (ed), *La estructura socieconómica de Castilla y León en la Unión Europea* (pp. 312–392). Centro de Documentación Europea (UVA), Valladolid.

Janhonen, M., & Johanson, J. E. (2011). Role of knowledge conversion and social networks in team performance. *International Journal of Information Management, 31*(3), 217–225.

Lucas, R. E. (1988). On the mechanics of economic development. *Journal of Monetary Economics, 22*(1), 3–42.

Lucas, R. E. (1993). Making a miracle. *Econometrica, 61*(2), 251–272.

McIver, D., Lengnick-Hall, C. A., Lengnick-Hall, M. L., & Ramachandran, T. (2012). Integrating knowledge and knowing: A framework for understanding knowledge in practice. *Human Resource Management Review, 22,* 86–99.

Mládková, L. (2012). Leadership in management of knowledge workers. *Procedia - Social and Behavioral Sciences, 41,* 243–250.

Nahapiet, J., & Ghoshal, S. (1998). Social capital, intellectual capital, and the organizational advantage. *Academy of Management Review, 23*(2), 242–266.

Nelson, R., & Winter, S. (1982). *An evolutionary theory of economic change*. Belknop Harvard.

Pajares, J., Hernández, C., & López, A. (2004). Modellin learning and R&D in innovative environments: A cognitive multi-agent approach. *JASSS. Journal of Artificial Societies and Social Simulation, 7*(2).

Pavit, K. (1990). What we know about the strategic management of technology. *California Management Review, 32*(3), 17–26.

Porter, M. E. (1990). *The competitive advantage of nations*. New York: Free Press.

Renzl, B. (2008). Trust in management and knowledge sharing: The mediating effects of fear and knowledge documentation. *Omega, 36,* 206–220.

Rifkin, J. (2014). *The zero marginal cost society: The internet of things, the collaborative commons, and the eclipse of capitalism*. Palgrave Macmillan.

Romer, P. (1986). Increasing return and long-run growth. *Journal of Political Economy, 94*(5), 1002–1037.

Rumelt, R. P., Schendel, D., & Teece, D. (1991). Strategic management and economics. *Strategic Management Journal, 12*(2), 5–29.

Sáiz-Bárcena, L., Díez, J. I., Manzanedo, M. A., & Rodríguez, C. (2013). Knowledge sharing in the company. *Learning from experience. Interciencia, 38*(8), 570–576.

Sáiz-Bárcena, L., Díez, J. I., Manzanedo, M. A., & del Olmo, R. (2014). Predisposition of workers to share knowledge: An empirical study. *Lecture Notes in Management and Industrial Engineering, 2,* 77–84.

Schoemaker, P. J. H. (1992). How to link strategic vision to core capabilities. *Sloan Management Review, 34*(1), 67–81.

Solow, R. (1956). A contribution to the theory of economic growth. *The Quarterly Journal of Economics, 70*(1), 65–94.

Chapter 9
Institutional Endowment and Innovation Strategies

Cesar Gamez-Alcalde, Carmen De Pablos-Heredero and Natalia Martin-Cruz

Abstract Firms understand that cooperation is key to sustainable innovation. Knowledge accumulated inside the firm requires external catharsis if the firm is to reinvent itself. Institutional endowment influences those cooperative processes, acting as a barrier or leverage. When building balanced cooperation strategies, the extent to which firms make their contributions counts in the innovation race by using "the wind behind" them, and it will determine how sustainable their performance is. Understanding the impact of institutional endowment on innovation strategies might reduce causal ambiguity and therefore help firms to contextualize their innovation strategies. The aim of this paper is to contribute with a type of cooperative innovation strategy dependent on two institutional factors: culture networking and the legal system.

Keywords Innovation · Institutional endowment · Country-specific context · Hofstede factors · Legal system

---

C. Gamez-Alcalde
Dpto. de Economía de La Empresa, Facultad de Ciencias Sociales y Jurídicas, Carlos III University, Campus de Getafe Calle Madrid, 126, 28903 Getafe Madrid, Spain
e-mail: cesar.gamez@uc3m.es

C. De Pablos-Heredero
Dpto. Economía de la Empresa (ADO), Economía Aplicada II y Fundamentos del Análisis Económico, Facultad de Ciencias Jurídicas y Sociales, Rey Juan Carlos University, Paseo Artilleros s/n, 28032 Vicalvaro Madrid, Spain
e-mail: carmen.depablos@urjc.es

N. Martin-Cruz (✉)
Grupo INSISOC, Dpto. de Organización de Empresas y CIM, Facultad de Ciencias Económicas y Empresariales, Universidad de Valladolid, Avda. Valle Esgueva 6, 47011 Valladolid, Spain
e-mail: ambiela@eco.uva.es

C. Hernández (ed.), *Advances in Management Engineering*, Lecture Notes in Management and Industrial Engineering, DOI 10.1007/978-3-319-55889-9_9

1 Introduction

Innovation is part of our DNA. This sentence is a classic in CEO statements to the media. Firms involved in every type industry invest large amounts of money in innovation (Jaruzelski et al. 2012). The latest report by Strategy and shows potential successful innovation strategies developed by firms in several countries: need-seekers, market-readers and technology-drivers (Jaruzelski et al. 2014). As to which factors prove key to achieving those strategies, they suggest having the right people in the right place and aligning the firm's strategy with its innovation strategy (vertical fit).

Some firms are examples of successful innovation strategies such as Apple under the managerial skills of Steve Jobs in the computer or cell phone industries (Martín García and De Pablos Heredero 2014). The other side of the coin is Kodak, a firm which lost its leadership in innovation in the photography industry. In 2009, its former CEO Antonio Perez, who remained head until 2013, suggested that the firm had lost its leadership in innovation in the photography sector not because it had ceased to invest in R&D but because it had been unable to exploit the technology in which it had been investing internally for years (interview with Antonio Pérez in the Waldorf Astoria, New York, Alumni Reunion IESE 2009). Both examples highlight the need to continue exploring which innovation strategies might prove successful or which might, on the other hand, spark the company's demise.

Firms such as Forbes, the Boston Consulting Group or Thompson Reuters de-vote their efforts to highlighting which firms are most innovative by utilizing various forms of analysis such as the Forbes Innovation Premium, the Boston Consulting Group innovation survey or the four criteria applied by Thompson Reuters. Year after year, the results prove overwhelming, since the majority of firms ranked highest are located in the US. For instance, in 2014, the results of the Boston Consulting Group showed that half of the most innovative firms are in the US and that there were no European or Latin American firms. According to the Forbes ranking, amongst the top ten, 60% of the most innovative companies are in the US. In its report, Thompson Reuters found that 39% of the most innovative firms came from Japan and 35% from the US; Europe accounted for 17% (with France and Switzerland heading the list).

Despite European Union attempts to promote innovation amongst its firms, as reflected in its reports (Koekoek 2012) and more recently in the Horizon (2020) programme which encourages particular effort with regard to SMEs, it seems that the desired results are not being achieved when compared to US or Japanese counterparts. In countries such as Spain, the Cotec report (2014) continues to reflect a certain pessimism vis-à-vis Spanish firms' capacity to innovate, bearing in mind the lack of sufficient support from public authorities, the absence of a financial market culture which funds investment, coupled with the fact that firms devote little in the way of financial and human resources to innovation (page 145). The opinions reflected in said report also seem to question the outcome of many of the initiatives undertaken to boost innovation in EU countries (2014 European Commission report on the state of innovation in the period 2010–2014–State of Innovation Union). These environments do, however, provide some examples of organizations that are capable of innovating, such as Real Madrid, which uses a two-fold innovation sports strategy (buying, loaning and training players) that allows it to achieve sustainable innovation (Blanco Callejo and Forcadell Martínez 2006).

There is ample research that seeks to unveil the process through which innovation takes place, both from the standpoint of industrial economics as well as through management theories such as the resource-based approach (Vega-Jurado et al. 2008). Of late, researchers seem to have taken an interest in the role played by internal and external knowledge in innovation, using theoretical approaches such as open innovation literature (Love et al. 2014) or the knowledge based approach (Herstad et al. 2015). A fresh line of inquiry, stemming from the progress made in open innovation, focuses on exploring hybrid innovation strategies which merge internal elements of knowledge with the knowledge provided by other stakeholders; in sum, collaborative innovative strategies (Herstad et al. 2014; Felin and Zenger 2014). In this regard, the literature addressing outsourcing allows additional elements to be incorporated into the study of business innovation processes.

The strategy of outsourcing business activities has had a major impact from a range of theoretical approaches, although it has mainly been the transaction cost theory (Williamson 1979, 1989, 1991) which has made the greatest contributions vis-à-vis pinpointing the most efficient ways of outsourcing transactions (Klein et al. 1978; Nooteboom 1993) and which has provided the greatest number of empirical studies aimed at demonstrating the theoretical claims (John and Weitz 1988; Jones and Hill 1988; Klein et al. 1990; Heide and Stump 1995). Adopting this theoretical approach, the variables identified for taking decisions are: asset specificity, uncertainty and frequency of transactions in situations of possible opportunism, incomplete information and limited rationality. For its part, the dynamic capabilities approach (Teece et al. 1997) has introduced the notion of dynamic capabilities which identify the optimal conditions in which long-term business relations, such as cooperation agreements, alliances, and other strategies can be maintained (Eisenhart and Martin 2000; Winter 2003). These elements include those related to knowledge management (KIW—knowledge internal workers) and change (strategic fit, potential conflicts between firms), as well as those related more specifically to human resources (cost/benefit for human resources by adopting a specific outsourcing strategy). More recently, the OLI approach (Ownership-Location-Internalization) posits other determinants in the outsourcing decision-making context which need to be taken into account (Martínez-Noya et al. 2010).

Therefore, the dilemma facing firms with regard to how to manage the innovation strategy that best fits a given institutional framework stands very much at the fore and further inquiry into the topic remains essential.

2 Firms Innovation Strategies in Open Contexts

The institutional framework or context has been posited as one of the key factors in a firm's capacity to innovate (Aldrich and Fiol 1994). However, determining which innovation strategies best fit the specific features of a firm in a given context is a topic that has yet to be explored in sufficient depth.

The literature addressing strategy has provided further insights into our understanding of the country-related factors which shape how firms act and the strategies they employ (Delios and Henisz 2003, 2000; Henisz and Delios 2001). A number of different arguments have been conjectured, such as political/regulatory (Kingley et al. 2012; Aragón-Correa et al. 2008), legal (Arruñada and Andonova 2008) or cultural (Hofstede 2009, 2010, 2012). Firms decide how to enter a country depending on the legal safety net the latter offers (Henisz and Zelner 2001, 2005; Delios and Henisz 2003). Specifically, firms' investment strategies in infrastructure are determined by the political structure and regulatory discretion in place in the various countries (Henisz and Zelner 2001). Political stability can also impact on international expansion strategy (Delios and Henisz 2003). Said study observes the ability of Japanese firms to manage political uncertainty in the country of entry through an investment sequencing strategy. In this regard, in countries displaying major political instability firms tend to opt for joint strategies rather than making direct investments. Analysis has also addressed firms' decisions with regard to their environmental strategy depending on the environmental restrictions imposed in each country (Aragón-Correa 1998; Aragón-Correa et al. 2008; Marcus et al. 2011). In their 2008 study, Aragón-Correa et al. explore firms' strategic responses to the issue of the environment in terms of the applicable regulations established by the country in which the firms operate. Possibilities range from a strategy of non-compliance to one of environmental leadership, which was shown to have a significant impact on business performance. As regards a country's cultural factors, Hofstede et al. (2010) point to different models of leadership that entail different types of decision-making strategies depending on the CEO's country of origin, with Swedish management tending to take risky decisions whilst at the same time also displaying a concern for the quality of the jobs in their firms. For other kinds of decision, Hofstede's work explores in depth the effect of a country's cultural factors on marketing strategies (Mooij and Hofstede 2010).

In the specific field of innovation strategies, the literature is progressing in its understanding of corporate strategies. With regard to the issue of outsourcing/offshoring R&D/innovation decisions, the importance of the institutional environment has been highlighted (Martínez-Noya et al. 2010; Martínez-Noya and García-Canal 2011). Specifically, variables such as labour cost structure, knowledge specialization or a country's intellectual property rights protection laws are key when choosing to outsource research activities (Martínez-Noya and García-Canal 2011; Hernández and Nieto 2015). The central role played by a country's culture has also been highlighted as crucial when developing strategic innovation alliances (Pothukuchi et al. 2002; Sirmon and Lane 2004). Is sum, institutional differences have proven their importance when innovation strategies are being worked out (Whitley 2000).

3 Innovation, Cultural Factors and Social Networks

Over time, the literature has evidenced how the cultural factors which characterize a country prove key to the creativity and innovation of firms located therein (Chua et al. 2015). Analysis has covered such wide ranging issues as a comparison between countries' cultural features and the levels of creativity in each (Lubart 1999; Saeki et al. 2001). It has also focused on the values which recognise the importance of collectivism, individualism or risk-aversion in individual creativity (Shane 1995; Hofstede 2001; Erez and Nouri 2010), and addressed the psychological approach, which centres on social norms as antecedents of the differences in creativity amongst individuals (Zou et al. 2009; Mok and Morris 2010).

From an organizational standpoint, it can be seen how creative environments engender more opportunities for firms to innovate (Chua et al. 2015). Indeed, firms implement knowledge seeking strategies overseas in an attempt to promote the innovation process (Almeida 1996; Frost 2001). Chua et al. (2015) propose a model of cultural alignment for overall creativity, considering two specific contextual variables—cultural distance and cultural 'oppression'—as determinants of successful innovative behaviour. In their model, these variables impact on the extent to which individuals commit to a creative activity and on the success of the innovation undertaken. In their model, cultural distance moderates the influence of cultural 'oppression' on individual creativity. One contribution said authors make is to include the role of cultural 'oppression' not only of the individual who is innovating but also of the target audience whom the innovation is aimed at. Such studies prove relevant in the sense that business initiatives and strategies might fail if employees have been trained in environments that display strong cultural 'oppression' and amongst whom there is no cultural distance. The analysis thus embraces the potential for cooperation between innovators from different backgrounds in terms of the variables pointed to previously. Environments of this nature stress the role of cooperation in innovation when seeking to improve innovation performance.

A complementary perspective is provided by Godart et al. (2012) who posit a new model of creative innovation based on international experience to gauge what impact managerial experiences overseas might have on creativity and business innovation. On a sample of eleven collections of haut couture, their study shows that firms which hire managers who have moderate levels of experience, integration and cultural distance in their professional experiences in various countries obtain the highest levels of creative innovation (Godart et al. 2012). Said finding highlights that firms wishing to generate innovation should devise a human resources policy which takes account of these aspects of the individual. In particular, a firm operating in relatively uncreative environments might opt for innovation strategies based on hiring managers who have wide international professional experience in several countries and who display a certain cultural distance with respect to the country in which the firm is located. A second option would be to cooperate with firms from other countries that evidence a certain cultural distance. According to the study, both innovation strategies might enable firms to secure creative and innovative performance.

Following this line of argument, the economic perspective of sociology underpins the role played by social networks in creativity and innovation (Ahuja 2000). Going one step further, Wang et al. (2014) include the role of knowledge networks in this relation. This progress in our understanding of the innovation process allows efficient innovation strategies to be identified. Specifically, said authors posit that the search for fresh knowledge will take place outside the firm in the extent to which its employees maintain a dense network of knowledge and provided there is not a high (or scant) centralization of knowledge. This finding means that, as they fail to maintain the right networks and centralized knowledge; innovative firms will require departments charged with seeking out new elements of external knowledge. In highly conventional and traditional contexts or environments it will thus prove essential to cooperate in order to innovate. Convincing a firm's employees to devote their time to exploring fresh knowledge will therefore only be possible if external cooperation networks are established.

4 Innovation Strategies and the Legal System

In a global world, innovative firms will decide their innovation strategy based on the characteristics of the legal system in place where they operate (Hall and Ziedonis 2001). Literature addressing the new institutional economics underscores that countries based on common law display greater copyright protection (La Porta et al. 1997, 1998). In any knowledge management process required to foster innovation, firms seek to protect the outcome thereof, leading them to implement different strategies depending on the protection offered by the legal system. Previous studies show how a weak copyright system encourages firms to engage in strategies centred on distribution rather than local production (Javorcik 2004). Changes in the copyright protection systems also impact on business strategies. In the case of the US, including new areas to be protected such as software or business models in the patent system in the late 90s has led to greater privatization of knowledge (Coriat and Orsi 2002).

Progress in the understanding of the role played by context in innovation still leaves much room to pinpoint which elements drive firms to opt for one particular innovation strategy. The recent work by Herstad et al. (2015) exploring in depth two modes of innovation ('science–technology–innovation' and 'doing–using–interacting') and by Chen and Adamson (2015) addressing the two paradigms of innovation ('random variation' and 'creative synthesis') provides further insights into the mechanisms linking the specific elements of context and innovation strategies. Both authors stress the importance of merging different modes or paradigms so as to achieve positive innovation outcomes, respectively. Both approaches concur vis-à-vis the role played by knowledge in achieving business innovation to propose a combination of modes/paradigms. The need to create routines of knowledge whilst encouraging a break with knowledge as a result of environmental dynamism underlies both formulations. Given that most firms operate in many contexts which display quite diverse characteristics in the various activities involved in their chain of value, the conclusions to emerge from these works prove key to understanding their 'ambiguous' innovation strategies.

The review of the literature leads us to draw a series of inferences concerning the relations between context and business innovation strategies. As has been highlighted, the two contextual variables which emerge as crucial to devising an innovation strategy are cultural (pro-innovation, no-innovation contexts) and legal (common law, civil law), in that both include key aspects which impact on the extent to which efforts in innovation prove successful. By contrasting the two aspects simultaneously when choosing an innovation strategy, different possibilities for firms emerge (Table 1).

Environments in which firms find themselves faced with a 'pro-innovation' culture and a legal system based on common law will tend to devise and implement cooperative innovation strategies based on relations with other firms, other actors that will support them when it comes to generating ideas, that ensure the creativity needed to maintain the levels of innovation required in the sectors/markets in which they operate. For their part, in 'no-innovation' contexts and legal systems based on civil law, the opposite situation will be true: non-cooperative innovation strategies. In other words, firms will develop their innovation strategies internally, using the resources and skills which they themselves are capable of generating. The two other situations emerge when the context is one of 'pro-innovation' and the legal system is based on common law or when the context is 'no-innovation' and the legal system is based on civil law. In these two instances, contextual factors represent forces that run contrary to the development of a univocal innovation strategy, which we refer to as 'ambiguous' innovation strategies. In other words, depending on the impact of one contextual factor or another, firms will tend to seek a balance between cooperation and internally developing innovation.

Table 1 Institutional endowment and firms strategies for innovation

Culture/legal system[a]	Common law	Civil law
Pro-innovation	Cooperative strategies for innovation	Ambiguous strategies for innovation
No-innovation	Ambiguous strategies for innovation	No-cooperative strategies for innovation

[a]Own source

5 Conclusion

Despite ever-increasing globalization, the role played by the context proves key when developing a firm's strategy and, in particular, its innovation strategy. The cultural and legal factors addressed in the present work affect how firms face up to innovation in its many different aspects, such as how staff is hired, incentives or the protection of ideas and how well the process performs. Literature addressing the impact these factors have on strategy is abundant although not systematized and it thus proves difficult to offer conclusive recommendations.

In innovation strategy, the decision concerning how to innovate, within the firm through internal investment, whether through purchasing external knowledge or in conjunction with other organisations, the context might provide companies with information that could help them make a good decision. A creative context which safeguards innovation will prove conducive towards helping firms devise cooperative innovation strategies, exchange knowledge and make best use of synergies with other firms. By contrast, a context which does not favour innovation and which offers scant mechanisms to protect property rights will make firms tend to consider and ultimately implement internal innovation strategies that protect the outcomes of innovation processes and which will allow them to devise mechanisms for managing knowledge which are appropriate for fostering creativity and securing the launch of new products onto the market in a sustainable manner. In between these two extremes lie 'hybrid' contexts which merge the two conditions of innovation, one creative, yet lacking the necessary mechanisms to protect innovation, one 'non-creative' yet with the assurance that the results of the innovation will be protected. Firms operating in these contexts will have to opt for 'ambiguous' innovation strategies, in other words, merging internal and external cooperative processes of innovation so as to benefit from one of the factors offered by the context whilst safeguarding themselves from the latter.

The arguments put forward concur with the results which systematically emerge when dealing with innovation in Europe, and which reflect that most innovative firms are located in countries such as Ireland or the northern countries, whereas countries such as Slovenia rank below the European Union average (Community Innovation Survey 2015).

In this regard, although globalization is a 19th century phenomenon (O'Rourke and Williamson 2002), in the matter of business strategies and, in particular, in innovation strategies, said phenomenon has had the greatest impact in the late 20th and early 21st. This is so much so that when developing their innovation strategy, the globalization of markets and supply, which affects the how firms' institutional framework develops, has enabled firms to adapt simultaneously to global processes or norms such as those related to quality (ISO standards were first created in 1926), the environment (ISO 14,000 was first published in 1996) or process appraisal (CMM was developed in 1986 and CMMI in 2001) that allow them to expand and penetrate any market. Domestic firms are now thinking globally like multinationals and innovation strategies are affected highly diverse institutional frameworks, which is reflected in a gradual evolution towards 'ambiguous' innovation models. As these global standards or models come to impose themselves, when devising their innovation strategy firms will need to evaluate a greater number of contextual factors and how they fit in with their available resources and capacities accumulated at each moment, in addition to which firms' innovation strategies will tend to resemble each other more.

In this sense, firms are implementing a business innovation model in the 21st century which admits that the protection rights afforded to their innovations is not perfect and that they will require more than mere patents if they are to have time to recoup the investments made to obtain a new product or service. In this model, firms aim to minimize the time to market, which entails getting as close to the market as possible (market-oriented innovation) through models such as open innovation with the consumer whilst seeking the more appropriate organizational forms to utilize their resources as efficiently as possible (technology-oriented innovation) and, if necessary, cooperating with partners so as to obtain new products or services.

References

Ahuja, G. (2000). Collaboration networks, structural holes, and innovation: A longitudinal study. *Administrative Science Quarterly, 45,* 425–455.

Aldrich, H. E., & Fiol, C. M. (1994). Fools rush in? The institutional context of industry creation. *Academy of Management Review, 19*(4), 645–670.

Almeida, P. (1996). Knowledge sourcing by foreign multinationals: Patent citation analysis in the U.S. semiconductor industry. *Strategic Management Journal, 17,* 155–165.

Aragón-Correa, J. A. (1998). Strategic proactivity and firm approach to the natural environment. *Academy of Management Journal, 41*(5), 556–567.

Aragón-Correa, J. A., Hurtado-Torres, N., Sharma, S., & García-Morales, V. J. (2008). Environmental strategy and performance in small firms: A resource-based perspective. *Journal of Environmental Management, 86*(1), 88–103.

Arruñada, B., & Andonova, V. (2008). Common law and civil law as Pro-market adaptations. *Washington University Journal of Law & Policy, 26,* 81–130.

Blanco Callejo, M. B., & Forcadell Martínez, F. J. F. (2006). El Real Madrid Club de Fútbol: la aplicación de un modelo empresarial a una entidad deportiva en España. *Universia Business Review, 3*(11), 36–61.

Chen, J., & Adamson, C. (2015). Innovation: Integration of random variation and creative synthesis. *Academy of Management Review, 40*(3), 461–473.

Chua R. Y. J., Roth, Y., & Lemoine, J. F. (2015). The impact of culture on creativity: How cultural tightness and cultural distance affect global innovation crowd sourcing work. Administrative Science Quarterly 602189227.

Coriat, B., & Orsi, F. (2002). Establishing a new intellectual property rights regime in the United States. Origins, content and problems. *Research Policy, 31,* 1491–1507.

Delios, A., & Henisz, W. (2003). Political hazards, experience and sequential entry strategies: The international expansion of Japanese firms, 1980–1998. *Strategic Management Journal, 24*(11), 1153–1164.

Delios, A., & Henisz, W. (2000). Japanese firm's investment strategies in emerging economies. *Academy of Management Journal, 43*(3), 305–323.

Eisenhardt, K. M., & Martin, J. A. (2000). Dynamic capabilities: What are they? *Strategic Management Journal, 21*(10–11), 1105–1121.

Pérez, A. (2009). *Entrevista en el hotel Waldorf Astoria*. Alumni Reunion IESE: Nueva York.

Erez, M., & Nouri, R. (2010). Creativity: The influence of cultural, social, and work contexts. *Management and Organization Review, 6*, 351–370.

Felin, T., & Zenger, T. R. (2014). Closed or open innovation? Problem solving and the governance choice. *Research Policy, 43*(5), 914–925.

Frost, T.S. (2001). The geographic sources of foreign subsidiaries' innovations. *Strategic Management Journal, 22,* 101–123.

Godart, F. C., Maddux, W. W., Shipilov, A. V., & Galinsky, A. D. (2012). Fashion with a foreign flair: Professional experiences abroad facilitate the creative innovations of organizations. *Academy of Management Journal, 58*(1), 195–220.

Hall, B.H., & Ziedonis, R. H. (2001). The patent paradox revisited: An empirical study of patenting in the US semiconductor industry, 1979–1995. RAND Journal of Economics spring (pp. 101–128).

Heide, J. B., & Stump, R. L. (1995). Performance implications of buyer-supplier relationships in industrial markets: A transaction cost explanation. *Journal of Business Research, 32*(1), 57–66.

Henisz, W., & Delios, A. (2001). Uncertainty, imitation and plant location: Japanese multinational corporations, 1990–96. *Administrative Science Quarterly, 46*(3), 443–475.

Henisz, W., & Zelner, B. (2001). The institutional environment for telecommunications investment. *Journal of Economics and Management Strategy, 10*(1), 123–148.

Henisz, Witold, & Zelner, B. (2005). Legitimacy, interest group pressures and change in emergent institutions: The case of foreign investors and host country governments. *Academy of Management Review, 30*(2), 361–382.

Hernández, V. & Nieto, M. J. (2015). Inward–outward connections and their impact on firm growth. International Business Review in press.

Herstad, S., Aslesen, H. W., & Ebersberger, B. (2014). On industrial knowledge bases, commercial opportunities and global innovation network linkages. *Research Policy, 43,* 495–504.

Herstad, S. J., Sandven, T., & Ebersberger, B. (2015). Recruitment, knowledge integration and modes of innovation. *Research Policy, 44,* 138–153.

Hofstede, G. (2001). Culture's recent consequences: Using dimension scores in theory and research. *International Journal of Cross-Cultural Management, 1*(1), 11–17.

Hofstede, G. (2009). American culture and the 2008 financial crisis. *European Business Review, 21*(4), 307–312.

Hofstede, G. (2010). The globe debate: Back to relevance. *Journal of International Business Studies, 41*(8), 1339–1346.

Hofstede, G., Garibaldi de Hilal, A. V., Malvezzi, S., Tanure, B., & Vinken, H. (2010). Comparing regional cultures within a country: Lessons from Brazil. *Journal of Cross-Cultural Psychology, 41*(3), 336–352.

Hofstede, G., & Minkov, M. (2012). Hofstede's fifth dimension: New evidence from the world values survey. *Journal of Cross-Cultural Psychology, 43*(1), 3–14.

Informe Cotec (2014). Impacto de la regulación sobre la innovación Colección Innovación Práctica, Fundación Cotec para la Innovación Tecnológica 66 p.

Horizon (2020). The EU Framework Programme for Research and Innovation. https://ec.europa.eu/programmes/horizon2020/en/h2020-section/innovation-smes

Jaruzelski, B., Loehr, J., & Holman, R. (2012). The global innovation 1000. Making ideas work. *Strategy+Business, Booz&Co winter, 69,* 1–17.

Jaruzelski, B., Staack, V., & Goehle, B. (2014). Proven paths to innovation success. Ten years of research reveal the best R&D strategies for the decade ahead. *Strategy+Business, Booz&Co winter, 77,* 1–18.

Javorcik, B. S. (2004). The composition of foreign direct investment and protection of intellectual property rights: Evidence from transition economies. *European Economic Review, 48*(1), 39–62.

John, G., & Weitz, B. A. (1988). Forward integration into distribution: An empirical test of transaction cost analysis. *Journal of Law Economics and Organization, 4*(2), 337–355.

Jones, G. R., & Hill, C. W. (1988). Transaction cost analysis of strategy-structure choice. *Strategic Management Journal, 9*(2), 159–172.

Kingley, A., Vanden Bergh, R., & Bonardi, J. P. (2012). Political market and regulatory uncertainty: Insights and implications for integrated strategy. *Academy of Management Perspectives, 26*(3), 52–67.

Klein, B., Crawford, R. G., & Alchian, A. A. (1978). Vertical integration, appropriable rents, and the competitive contracting process. *Journal of Law and Economics, 21*(2), 297–326.

Klein, S., Frazier, G. L., & Roth, V. J. (1990). A transaction cost analysis model of channel integration in international markets. *Journal of Marketing Research, 27,* 196–208.

Koekoek, P. (2012). Innovative companies: Making Europe the best place to grow highly innovative ICT SMEs. Report of a High-Level Panel Discussion. DG Connect Brussels 17 September 2012.

La Porta, R., Lopez-de-Silanes, F., Shleifer, A., & Vishny, R. W. (1997). Legal determinants of external finance. *Journal of Finance, 52*(3), 1131–1150.

La Porta, R., Lopez-de-Silanes, F., Shleifer, A., & Vishny, R. W. (1998). Agency problems and dividend policies around the world. National Bureau of Economic Research w6594.

Love, J. H., Roper, S., & Vahter, P. (2014). Dynamic complementarities in innovation strategies. *Research Policy, 43,* 1774–1784.

Lubart, T. I. (1999). Creativity across cultures. In R. J. Sternberg (Ed) *Handbook of creativity* (pp. 339–350). New York: Cambridge University Press.

Marcus, A., Aragon-Correa, J. A., & Pinkse, J. (2011). Firms, regulatory uncertainty, and the natural environment. *California Management Review, 54*(1), 5–16.

Martín García, P., & de Pablos, Heredero C. (2014). Liderazgo acomodaticio y resultados empresariales: Una aproximación desde el análisis de Steve Jobs. *Interciencia, 39*(9), 673–680.

Martínez-Noya A, García-Canal, E., & Guillén, M. F. (2010). Why do firms locate R&D outsourcing agreements offshore? The role of ownership, location, and externalization advantages. Fundación de las Cajas de Ahorros. Documento de trabajo nº 526/2010:1–58.

Martínez-Noya, A., & García-Canal, E. (2011). Technological capabilities and the decision to outsource/outsource offshore R&D services. *International Business Review, 20*(3), 264–277.

Mok, A., & Morris, M. W. (2010). Asian-Americans' creative styles in Asian and American situations: Assimilative and contrastive responses as a function of bicultural identity integration. *Management and Organization Review, 6,* 371–390.

Mooij, M., & Hofstede, G. (2010). Applications to global branding and advertising strategy and research. *International Journal of Advertising: The Review of Marketing Communications, 29* (1), 85–110.

Nooteboom, B. (1993). Firm size effects on transaction costs. *Small Business Economics, 5*(4), 283–295.

O'Rourke, K. H., & Williamson, J. G. (2002). When did globalisation begin? *European Review of Economic History, 6*(1), 23–50.

Pothukuchi, V., Damanpour, F., Choi, J., Chen, C. C., & Park, S. H. (2002). National and organizational culture differences and international joint venture performance. *Journal of International Business Studies, 33,* 243–265.

Saeki, N., Fan, X., & Dusen, L. (2001). A comparative study of creative thinking of American and Japanese college students. *The Journal of Creative Behavior, 35*(1), 24–36.

Shane, S. (1995). Uncertainty avoidance and the preference for innovation championing roles. *Journal of International Business Studies, 26,* 47–68.

Sirmon, D. G., & Lane, P. J. (2004). A model of cultural differences and international alliance performance. *Journal of International Business Studies, 35*(4), 306–319.

Teece, D. J., Pisano, G., & Shuen, A. (1997). Dynamic capabilities and strategic management. *Strategic Management Journal, 18*(7), 509–533.

Vega-Jurado, J., Gutiérrez-Gracia, A., Fernández-de-Lucio, I., & Manjarrés-Henríquez, L. (2008). The effect of external and internal factors on firms' product innovation. *Research Policy, 37*(4), 616–632.
Wang, C., Rodan, S., Fruin, M., & Xu, X. (2014). Knowledge networks, collaboration networks, and exploratory innovation. *Academy of Management Journal, 57*(2), 484–514.
Whitley, R. (2000). The institutional structuring of innovation strategies: business systems, firm types and patterns of technical change in different market economies. *Organization Studies, 21* (5), 855–886.
Williamson, O. E. (1979). Transaction-cost economics: The governance of contractual relations. *Journal of Law and Economics, 2,* 233–261.
Williamson, O. E. (1989). Transactions cost economics. In R. Willig (Eds) *Handbook of industrial organization* vol. 1 North-Holland (pp. 135–182).
Williamson, O. E. (1991). Comparative economic organization: the analysis of discrete structural alternatives. *Administrative Science Quarterly, 36,* 269–296.
Winter, S. G. (2003). Understanding dynamic capabilities. *Strategic Management Journal, 24*(10), 991–995.
Zou, X., Tam, K. P., Morris, M. W., Lee, S., Lau, I. Y. M., & Chiu, C. (2009). Culture as common sense: Perceived consensus versus personal beliefs as mechanisms of cultural influence. *Journal of Personality and Social Psychology, 97,* 579–597.

Part III
Heuristics, Intelligent Systems and Agent Based Modelling

Chapter 10
A Critical Sight to the NEH Heuristic

Ramón Companys and Imma Ribas

Abstract This paper analyses the behaviour of the insertion procedure of the NEH heuristics due to in some cases the procedure does not improve the solution quality. This fact is observed specially in the NEH-based heuristics that have been proposed in the literature of the blocking flow shop problem. As a result of this work, we recommend to evaluate the sequence before and after the insertion phase in order to retain the best of both.

Keywords Scheduling · Permutation flowshop · Blocking flowshop

1 Introduction

This work is part of a study which aims to analyse, in several variants of the flow shop scheduling problem, the reasons why the adaptation of the insertion phase of the NEH heuristic (Nawaz et al. 1983), do not always improve the sequence that it tries to improve. This is especially noticeable when the buffers between two stages are reduced to zero, known as blocking case. Interestingly many authors seem to ignore this fact and they still use this heuristics in order to find good initial solutions for their algorithms.

Many simple constructive heuristics used to solve the permutation flow shop scheduling problem (PFSP) build a sequence of the n jobs by n iterations, in which in each one an unscheduled job is selected and is put in a relative o absolute

R. Companys (✉)
Departament d'Organització d'Empreses, DOE—EPSEB—Universitat Politècnica de Catalunya, BarcelonaTech, Gregorio Marañón 44–50, 3rd Floor, 08028 Barcelona, Spain
e-mail: ramon.companys@upc.edu

I. Ribas
Departament d'Organització d'Empreses, DOE—ETSEIB—Universitat Politècnica de Catalunya, BarcelonaTech, Avda. Diagonal, 647, 7th Floor, 08028 Barcelona, Spain
e-mail: imma.ribas@upc.edu

C. Hernández (ed.), *Advances in Management Engineering*, Lecture Notes in Management and Industrial Engineering, DOI 10.1007/978-3-319-55889-9_10

position of the sequence. Absolute position means a fix position whereas a relative position means the precedence or succession respect to others jobs already scheduled. On the other hand, we refer to simple heuristic to those built with only two phases without improvement phase or other elements as the randomness or backtracking. In this research, the second phase of the simple heuristics considered is the insertion phase of NEH.

Some examples of this heuristics are Palmer (1965), Trapezium (Companys 1966), Gupta (1971), RA (Dannenbring 1977), etc. Therefore, at each iteration, two questions are answered:

Q1: Which job has to be scheduled?
Q2: In which position has to be scheduled?

The answer to the first question is found in the first phase by means a certain criterion, materialized in a static or dynamic index.

2 Problem Definition

A set of n jobs have to be processed in m machines. The route of jobs in all machines is the same, which allows the machines to be enumerated in the order defined by the route. Therefore, the first operation is in machine 1 and the last in machine m. Each job i, $i \in \{1, 2, \ldots, n\}$ requires a processing time p_{ij} in each machine j, $j \in \{1, 2, \ldots, m\}$. Machines and jobs are ready at instant zero. The setup times are considered to be included in the processing time. The objective is to find a sequence of jobs to be processed such that the makespan is minimized.

According to the notation proposed by Graham et al. (1979) the problem can be noted as $Fm|prmu|C_{max}$. In these problems, a solution is formed by one of the $n!$ possible permutation of jobs.

Alternatively, some industrial systems have to be modelled as a flowshop without buffers between two consecutive machines. In these cases, the jobs cannot leave the machine until the next machine is free, which might cause the blockage of the machine. This problem is denoted as $Fm|block|C_{max}$.

Given the processing time of jobs p_{ij} and a sequence σ of the n jobs, a set of recurrent equation allows calculating the makespan associate to the active program either of the $Fm|prmu|C_{max}$ and $Fm|block|C_{max}$. Such equation can be found in Ribas et al. (2011, 2013).

3 The NEH Algorithm

The NEH heuristic was designed for the $Fm|prmu|C_{max}$ problem. This procedure gave better solution than other heuristics methods but with greater amount of work albeit lower than the exact procedures based in branch-and-bound, as the proposed by Lomnicki (1965).

The NEH procedure can be divided into two steps:

Step 1: ordering jobs according the LPT rule.
Step 2: in accordance with the order established in step 1, take the first two jobs and schedule them in such a way that they minimize the partial makespan, considering an instance with only two jobs. Then for $k = 3$ up to n, insert the k-th job into one of the possible k positions of the partial sequence. The objective is to minimize the C_{max} of the $Fm|prmu|C_{max}$ problem with k jobs.

When step 2 is finished a new sequence of the n jobs is obtained, which for the problem considered, is normally a good solution. Step 1 allows answering to the first question whereas step2 to answers Q2.

The only difficulty is that in the classical implementation of the algorithm $[n \cdot (n + 1)/2 + 1] - 1$ programs must be calculated, of which n of them correspond to complete sequences. This makes the complexity of NEH reach to $O(n^3 \cdot m)$ which can be costly for large instances of the problem. However, Taillard (1990) presented an organization of calculation that reduces the complexity to $O(n^2 \cdot m)$. Surprisingly, the process is known as "Taillard's acceleration" rather than "Taillard's implementation". Obviously, the reduction of calculations and computational time has a cost because it is obtained less information than in other implementations.

It should be added that this organization is viable only if the criterion used to decide the best position to insert a job is the makespan. If the criterion is the total flowtime or total tardiness of jobs the acceleration cannot be implemented because it is necessary to know all the C_i values in order to decide the position to insert the job.

3.1 *Interpretation of Step 2*

We consider the insertion procedure as the fundamental characteristic of the NEH algorithm. The insertion procedure can be interpreted as a greedy algorithm and as well as a mapping. As greedy algorithm, it constructs the sequence by placing the jobs one by one, always in the "best" relative position given the partial sequence already built.

The final value of the objective function associated with the sequence is known only once the procedure is finished because it is not strictly the sum of the contributions of each individual job. Therefore, it is used an objective function, similar to the original, as an indicator to define the "best" position but restricted to the partial sequence consisting of the jobs already scheduled (which are the result of the order established in phase 1).

In the problem Fm|prmu|C_{max}, wherein the NEH algorithm works reasonably well, each job makes a contribution to the global objective function, and this contribution can be assessed more or less, by the impact of that job to the partial sequence. Moreover, to consider the order LPT for the initial sequence σ_0 is reasonable since it seems a good strategy to integrate jobs to the partial sequence so that the contribution of the first ones to the objective be potentially more important than the last ones, and $P_i = \sum_{J=1}^{m} p_{i,j}$ is a good index for that.

Alternatively, the insertion procedure can be interpreted as a mapping φ that from a sequence σ_0 obtains another sequence $\hat{\sigma} : \hat{\sigma} = j(\sigma_0)$. In the case that the sequence σ_0 be LPT its image usually have a better evaluation $C_{max}(\hat{\sigma}) < C_{max}(\sigma_0)$, but, as it will be shown, this property is not generalizable to all sequences $\sigma_0 \in \sum$.

Framinan and Leisten (2003) tested 22 indexes to build the sequence $\sigma_{0.}$ associated to 8 different ordering criterion, which lead to 176 initial sequences. They showed the obtained results for 2000 instances, with 4 values of *n* (25, 50, 75 and 100) and 5 of *m* (5, 10, 15, 20 and 25) and they deducted that the order LPT lead to the best results after applying the insertion procedure.

Nagano and Moccellin (2002) proposed an initial sequence, NM, not included among the 176 tested by Framinan et al. (2003), which gives some slightly better results, according to the authors. These better results were not confirmed on the Taillard's instances as (Kalczynski and Kamburowski 2008) already stated. Dong et al. (2008) also proposed a new ordering method, DCH, and (Kalczynski and Kamburowski 2008) another, KK, both are different from the 176 mentioned above.

Really, the sequences NM, KK and DCH represent minor corrections to the LPT rule. The use of the insertion procedure with this rules lead to disconcerting results. Denoting to the sequences obtained after the insertion Nlpt, Nnm, Ndhc and Nkk, only Ndch has a certain advantage over Nlpt, although this advantage, in relative values, is about 0.3%.

3.2 Ties

Ties occur in both phases of the NEH algorithm. Step 1 consist of establishing an initial sequence of jobs, namely ordering jobs according to the non-decreasing order of $P_i = \sum_{J=1}^{m} P_{i,j}$. What happens if two jobs *h* and *i* have the same sum of processing times, i.e. $P_i = P_h$? For example, in the set of Taillard's instances with $n = 500$ and $m = 20$, the number of different values of P_i is potentially 500, number which would be reached if all P_i values are different. The 1961 possible values are largely sufficient to none of the 500 jobs of an instance repeats value. However, none of the 10 instances of the set reaches to the 500 different values; the maximum number of different values is 328, for the TAIL0119. In TAIL0111, in which the number of different P_i values is 304 there exist many sequences compatibles with the order LPT, namely 1.53×10^{76}. It would be far to assume that all of them lead, after the step 2, to a permutation of the same quality. This is, to some extent, easy to check. According to our calculations the difference between the best and worst values can reach, in relative value, to 1.4%. In published papers in prestigious journals usually appear statements in which the authors claim the superiority of one method over another (for instance DCH against LPT) based on lower differences that the one that can cause a renumbering of the jobs, whose superiority comes endorsed by obscure statistical procedures. Therefore, the effect of renumbering cannot be considered trivial. How to choose the best LPT sequence? Or at least, one that be good? In the absence of specific tiebreaker rule acts the implicit rule and the obtained result depends on the order in which the jobs appears, that is, of their initial numbers. If these jobs had been described in a different order would lead to another solution, which in principle does not seem very reasonable (the results of a deterministic algorithm should be independent of the order in which are communicated the different objects that it manipulates). All that said for the LPT order applies to other rules.

This topic has received little attention in the literature, however the ties between two or more positions in the insertion phase, which have a lower impact, have been treated by Dong et al. (2008), Fernandez-Viagas and Framinan (2014), Kalczynski and Kamburowski (2008), Ribas et al. (2011).

3.3 *Is the Insertion Step Efficient as a Greedy Algorithm?*

Nagano and Moccellin (2002) already indicated that if the obtained sequence by NEH was treated by the insertion procedure not always lead to a better solution. By simulation, with the values of n and m indicated and with a sample of 100 instances of each (n, m) combination we have determined that from the LPT sequence, 99.964% of the sequences improve the first time that they are treated by the insertion method, but, after treating this sequences again by the insertion procedure, we saw that 88.607% of the sequences were worse, 4.321% obtained the same and, therefore, only 7.072% were better. If we establish a recurrent use of the insertion procedure from the sequence obtained by the LPT rule, after the fourth use, we observed that only in 19.071% of the instances some sequences were better than the obtained after the first use. The improvement of the sequences by the insertion procedure not even occurs with all LPT sequences, and presumably neither with all sequences NM, DHC and KK. Although the phenomenon is rare, we have collected some instances in which the evaluation of the LPT sequence is equal to or less than the sequence NEHlpt. Surprisingly, (Framinan and Leisten 2003) did not detected this phenomenon in their extensive experimentation.

4 NEH Algorithm in Fm|block|C_{max}

The Fm|block|C_{max} scheduling problem is the closed case to the Fm|prmu|C_{max}, therefore, the classic NEH algorithm were expected, once modified the calculations of the partial or global makespan of a sequence, due to possible blockings, it should show a good behaviour. In this sense, many authors state that (Leisten 1990) showed that the heuristic Nlpt is also efficient for this problem. Such authors should reread the work of Leisten since we think that he does not say exactly that. Ronconi (2004) did not find sufficiently effective the initial sequence LPT and proposed two different rules to replace:

- MinMax (MM). In this case, the sequence obtained after the insertion procedure is denoted as Nmm.
- PF, based on the profile fitting proposed by McCormick et al. (1989). The sequence obtained after the insertion procedure is denoted as Npf.

Later, (Pan and Wang 2012) proposed several variants of PF of which we have retained two, plus the original: wPF and PW. We denoted the sequences of such variants once have been processed by the insertion procedure, as Nwfp and Npw. One of the weak points of the procedures based on *profile fitting* consists in choosing the first job. Usually, the job with less P_i is chosen even though some authors test several jobs, normally the firsts in the SPT order, and they keep those that have given the best results, procedure that do not enter into our consideration. The choice of the job with less P_i is inconsistent with the rest of the steps in the procedures based on *profile fitting*. We put as the first job the one that would result from applying the procedure, assuming the existence of a previous job in position 0, such that $p_{j,0} = 0$ for all j.

5 Computational Evaluation

The computational evaluation has been done with the Taillard's instances and with instances generated ad hoc. The index used to evaluate the quality of solutions was the relative percentage deviation (RPD) of the obtained solution against the best solution known calculated as (1):

$$\mathrm{RPD} = \frac{\mathrm{Heur}_{h,s} - \mathrm{Best}_s}{\mathrm{Best}_s} \cdot 100 \tag{1}$$

where $\mathrm{Heur}_{h,s}$ is the C_{max} value obtained in instance s by heuristic h and Best_s is the best solution known for instance s. The best values for Taillard instances used for the Fm|block|C_{max} can be found in Ribas et al. (2013).

The instances generated ad hoc were created similar to the Taillard's. We generated 100 instances for each combination of $n \times m$ values were n = 25, 50,

100, 150, 200, 300 and 400 jobs and $m = 5$, 10, 15 and 20 machines, i.e. 2800 instances. In this case, $Best_s$ is the best value obtained by any heuristic. The next tables show the average relative percentage deviation (ARPD) per set in order to summarize the results.

Table 1 shows the Average RPD (ARPD) for each heuristic and set of Taillard's problems. We have added the Nspt procedure to the group of heuristics mentioned in this text because we have observed that with the blocking constraint the SPT sequence normally lead to better results than the LPT, as it can also be observed in this table. The "+" after the name of heuristics indicates that we have retained the best solution between the obtained before and after processing the sequence by the insertion procedure. Notice that we have omitted Nlpt+ and Nspt+ because, in all instances, the insertion procedure improves the sequence. However, we can observe differences in the other heuristics. In particular, in only one instance the sequence mm is better than Nmm but in the PF-based heuristics (Npf, Nwpf and Npw) the initial sequences are better in about 50%. Notice that the lower ARPD value is obtained by Npw+.

To have a broader perspective of the results, we have repeated this experiment with the 2800 instances generated ad hoc. Table 2 shows not only the ARPD values by each heuristic and set but also (last fourth columns) the percentage of instances in which the initial sequences is better than the sequence obtained after the insertion procedure. We have omitted columns LPT and SPT because all instances are improved by the insertion procedure. These results confirm the previous ones. The best constructive procedure is the Npw+ and it is advisable to evaluate the sequence before and after the insertion procedure to keep the best of two sequence, especially in PF-based heuristics because the insertion procedure worsen the sequence in about 50% of the instances.

Table 1 Average RPD values obtained by each heuristics for each set of Taillard's problems

$n \times m$	Nlpt	Nspt	Nmm	Nmm +	Npf	Npf+	Nwpf	Nwpf +	Npw	Npw +
20 × 5	5.628	6.056	5.434	5.434	5.255	5.255	6.100	5.393	5.768	5.054
20 × 10	5.441	5.522	5.605	5.605	5.513	5.478	5.696	5.592	5.146	4.955
20 × 20	3.429	5.709	3.642	3.642	5.599	5.599	5.300	5.300	5.193	5.147
50 × 5	8.321	6.754	6.561	6.561	6.991	6.475	7.255	7.006	7.523	7.144
50 × 10	8.713	7.617	6.888	6.888	7.391	7.391	7.634	7.629	8.070	7.874
50 × 20	7.759	6.520	6.260	6.260	6.869	6.869	6.433	6.433	6.319	6.319
100 × 5	9.631	8.453	7.908	7.834	7.990	6.250	7.723	5.904	7.817	6.155
100 × 10	9.216	7.267	6.911	6.911	7.287	6.499	7.839	6.649	8.040	6.084
100 × 20	7.086	6.145	5.924	5.924	6.198	6.026	6.117	6.059	6.111	6.008
200 × 10	8.621	6.955	7.097	7.097	7.042	4.143	7.859	4.698	7.764	4.604
200 × 20	7.497	5.781	5.617	5.617	5.767	4.967	6.151	4.449	5.936	4.451
500 × 20	7.084	5.882	5.724	5.724	5.774	2.662	6.336	2.751	6.350	2.718
All	7.369	6.555	6.131	6.125	6.477	5.616	6.705	5.630	6.672	5.525

Table 2 Average RPD values obtained by each heuristic and set generated ad hoc, and percentage of instances that the initial sequence is better

n x *m*	Nlpt	Nspt	Nmm+	Npf+	Nwpf+	Npw+	MM	PF	wPF	PW
25 × 5	2.248	1.772	1.961	1.340	1.544	1.562	22	10	4	10
25 × 10	1.676	1.493	1.566	1.770	1.534	1.348	0	0	1	10
25 × 15	1.042	1.747	1.326	1.655	1.579	1.609	0	0	2	0
25 × 20	0.954	1.950	1.261	2.015	1.634	1.557	0	0	2	0
50 × 5	2.668	1.628	1.414	1.104	1.123	0.831	2	30	42	50
50 × 10	1.538	0.954	0.954	0.924	1.083	1.023	0	10	6	10
50 × 15	1.305	0.985	0.693	1.123	1.096	1.039	0	0	0	0
50 × 20	0.931	1.117	0.917	1.295	1.110	0.994	0	0	1	0
100 × 5	3.833	2.749	2.360	0.526	1.092	0.798	0	90	92	90
100 × 10	2.369	1.307	1.233	0.679	0.811	0.642	0	60	67	70
100 × 15	1.422	0.696	0.706	0.719	0.775	0.600	0	20	33	40
100 × 20	1.117	0.601	0.605	0.768	0.666	0.538	0	0	6	10
150 × 5	5.103	4.015	3.316	0.565	0.778	0.503	7	100	100	100
150 × 10	3.728	2.655	2.494	0.513	0.893	0.651	0	100	100	100
150 × 15	2.245	1.179	1.220	0.736	0.674	0.540	0	70	90	90
150 × 20	1.432	0.650	0.663	0.748	0.608	0.510	0	20	57	70
200 × 5	6.333	5.109	4.274	0.378	0.702	0.522	10	100	100	100
200 × 10	4.680	3.458	3.226	0.398	0.787	0.656	0	100	100	100
200 × 15	3.222	2.122	2.125	0.638	0.617	0.484	0	90	100	100
200 × 20	2.004	1.189	1.245	0.829	0.381	0.259	0	60	93	100
300 × 5	7.242	6.208	4.833	0.292	0.690	0.444	54	100	100	100
300 × 10	5.957	4.646	4.357	0.178	0.847	0.648	0	100	100	100
300 × 15	4.404	3.212	3.239	0.417	0.471	0.409	0	100	100	100
300 × 20	3.328	2.291	2.279	0.738	0.392	0.280	0	100	100	100
400 × 5	8.153	6.964	5.112	0.233	0.591	0.424	90	100	100	100
400 × 10	6.712	5.472	5.197	0.123	0.765	0.751	0	100	100	100
400 × 15	5.279	4.109	4.000	0.221	0.640	0.583	0	100	100	100
400 × 20	4.094	3.067	3.019	0.442	0.351	0.232	0	100	100	100
All	3.394	2.619	2.343	0.763	0.865	0.730	6	59	64	67

6 Conclusions

This research demonstrates that if the procedures built with the adaptation of the insertion procedure (second step of NEH heuristic) are used to solve Fm|block|C_{max}, it is necessary to evaluate the sequence before and after the insertion step because this last step not always permits improving the previous sequence.

This research continues with other problems as Fm|prmu|ΣC_i, Fm|prmu|ΣT_i, Fm|block|ΣC_i and Fm|block|ΣT_i in which this effect is similar and, in some cases, even increased.

References

Companys, R. (1966). Métodos heurísticos en la resolución del problema del taller mecánic. *Estudios Empresariales, 5*(2), 7–18.

Dannenbring, D. G. (1977). An evaluation of flow shop sequencing heuristics. *Management Science, 23*(1), 1174–1182.

Dong, X., Huang, H., & Chen, P. (2008). An improved NEH-based heuristic for the permutation flowshop problem. *Part Special Issue: Telecommunications Network Engineering, 35*(12), 3962–3968.

Fernandez-Viagas, V., & Framinan, J. M. (2014). A bounded-search iterated greedy algorithm for the distributed permutation flowshop scheduling problem. *International Journal of Production Research, 53*(4), 1111–1123.

Framinan, J. M., & Leisten, R. (2003). An efficient constructive heuristic for flowtime minimisation in permutation flow shops. *Omega, 31*(4), 311–317.

Graham, R. L., Lawler, E. L., Lenstra, J. K., & Rinnooy Kan, A. H. G. (1979). Optimization and approximation in deterministic sequencing and scheduling: A survey. *Annals of Discrete Mathematics, 5,* 287–326.

Gupta, J. N. D. (1971). A functional heuristic algorithm for the flow shop scheduling problem. *Operational Research Quarterly, 22*(1), 39–47.

Kalczynski, P. J., & Kamburowski, J. (2008). An improved NEH heuristic to minimize makespan in permutation flow shops. *Part Special Issue: Bio-Inspired Methods in Combinatorial Optimization, 35*(9), 3001–3008.

Leisten, R. (1990). Flowshop sequencing problems with limited buffer storage. *International Journal of Production Research, 28*(11), 2085.

Lomnicki, Z. A. (1965). A branch and bound algorithm for the exact solution of three machine scheduling problems. *Operational Research Quarterly, 16*(1), 89–100.

McCormick, S. T., Pinedo, M. L., Shenker, S., & Wolf, B. (1989). Sequencing in an Assembly Line with Blocking to Minimize Cycle Time. *Operations Research, 37,* 925–936.

Nagano, M. S., & Moccellin, J. V. (2002). A high quality constructive heuristic for flow shop sequencing. *Journal of the Operational Research Society, 53,* 1374–1379.

Nawaz, M., Enscore, E. E., Jr., & Ham, I. (1983). A heuristic algorithm for the m-machine, n-job flow-shop sequencing problem. *Omega, 11*(1), 91–95.

Palmer, D. S. (1965). Sequencing jobs through a multi-stage process in the minimum total time-a quick method of obtaining a near optimum. *Operational Research Quarterly, 16,* 101–107.

Pan, Q.-K., & Wang, L. (2012). Effective heuristics for the blocking flowshop scheduling problem with makespan minimization. *Omega, 40*(2), 218–229.

Ribas, I., Companys, R., & Tort-Martorell, X. (2011). An iterated greedy algorithm for the flowshop scheduling problem with blocking. *Omega, 39*(3), 293–301.

Ribas, I., Companys, R., & Tort-Martorell, X. (2013). A competitive variable neighbourhood search algorithm for the blocking flowshop problem. *European J. of Industrial Engineering, 7* (6), 729–754.

Ronconi, D. P. (2004). A note on constructive heuristics for the flowshop problem with blocking. *International Journal of Production Economics, 87*(1), 39–48.

Taillard, E. (1990). Some efficient heuristic methods for the flow shop sequencing problem. *European Journal of Operational Research, 47*(1), 65–74.

Chapter 11
A Brief Introduction to the Use of Machine Learning Techniques in the Analysis of Agent-Based Models

María Pereda, José Ignacio Santos and José Manuel Galán

Abstract In this paper, we give a succinct introduction to some basic concepts imported from the fields of Machine and Statistical Learning that can be useful in the analysis of complex agent-based models (ABM). The paper presents some guidelines in the design of experiments. It then focuses on considering an ABM simulation as a computational experiment relating parameters with a response variable of interest, i.e. a statistic obtained from the simulation. This perspective gives the opportunity of using a supervised learning algorithm to fit the response with the parameters. The fitted model can be used to better interpret and understand the relation between the parameters of the ABM and the results in the simulation.

Keywords Agent based modelling · Machine learning · Simulation · Permutation test · Statistical learning

1 Agent-Based Modelling

Agent-based modelling (ABM) is currently one of the most active modelling paradigms in many scientific disciplines ranging from Sociology (Macy and Willer 2002) to Industrial Organization (Chang 2011) or Economics (Hernández et al. 2014). The gist of the approach lies on the particular process used to build the abstraction from the target system that is being studied. In an ABM model each

M. Pereda (✉) · J.I. Santos · J.M. Galán
Grupo INSISOC, Área de Organización de Empresas,
Dpto. de Ingeniería Civil, Escuela Politécnica Superior,
Universidad de Burgos, C/Villadiego S/N, 09001 Burgos, Spain
e-mail: mpereda@ubu.es

J.I. Santos
e-mail: jisantos@ubu.es

J.M. Galán
e-mail: jmgalan@ubu.es

C. Hernández (ed.), *Advances in Management Engineering*, Lecture Notes in Management and Industrial Engineering, DOI 10.1007/978-3-319-55889-9_11

entity identified in the target system is explicitly and individually represented as an agent, and the different interactions among the agents and the environment are also explicitly represented in the model. This direct correspondence provides the modeller with several interesting features, most of which are a consequence of making the abstraction process easier (Galán et al. 2009). In an ABM model it is almost straightforward to remove simplifying assumptions often used in other modelling paradigms, and consider the effect of heterogeneity, spatial influence, finite populations or bounded rationality, just to mention some examples.

This advantage at the modelling stage often increases the difficulty of analysis of the model, which is sometimes so complicated that it is not easy to understand the combined effects of all its assumptions. In the case of models used to illustrate a general mechanism or an emergent stylized fact, this circumstance is usually circumvented, if possible, obtaining closed analytical solutions, or exploring the complete range of parameter combinations by simulation. However, in models trying to reproduce specific and detailed situations, occasionally it is not easy to decide a priori those assumptions that should be simplified and those that are key elements to keep the model descriptive in terms of the analysed target. Analysts then face models with a number of parameters so large that a prohibitive quantity of computational resources is required to fully explore them, and so complicated that general analytical solutions are difficult to obtain.

The aim of this paper is to discuss a set of concepts and activities used in machine and statistical learning that can help to understand the behaviour of complicated (and not only complex) agent-based models. The rest of this work is structured as follows: The following section succinctly discusses how to sample a model in an efficient manner. We then explain why it can be useful to think about the results of a model as a classification or regression problem, and present possible avenues that an analyst can follow to adjust and interpret the model. Subsequently, a common way to analyse variable significance is discussed, and finally, conclusions are presented in the last section.

2 Design of Experiments with Space Filling Properties

An experiment is a procedure in which the input variables of a model (i.e. the system under study) are changed in order to analyse the reasons for the change in the *response variable* (a.k.a. output variable). The conduction of formal planned experimentation, i.e. Design of Experiments (DOE), is a crucial step that consists in getting the maximum amount of information from the model with the minimum amount of resources (the more samples, the more CPU time). This is carried out by properly choosing samples in the *design space* (part of the parameter space under study).

Given a parameter space in an ABM model, there are different ways of selecting samples and design experiments (Lee et al. 2015). Some of them have a particular configuration, such as Factorial Design, Central Composite, Taguchi, among others, consisting in discretising the parameter ranges in levels and sampling these values. Other approaches are the so-called space filling techniques, which aim to cover the design space uniformly and are based on statistical sampling.

When exploring a stochastic model, deciding whether to spend resources on exploring more diverse parameter combinations or replicating several times a given combination of parameters to reduce the uncertainty about the expected output value implies a trade-off that should be balanced and that is case-dependent.

The most obvious sampling technique is the Monte Carlo random sampling, which consists in sampling each parameter range randomly. The problem with this technique is that the design space is not covered evenly, but there can appear clusters of samples and empty spaces by chance. Another common issue while planning a DOE for computer models is that, sometimes, large design spaces need to be explored.

A space filling sampling technique that enjoys great popularity in computer simulation is Latin Hypercube Sampling (LHS) (McKay et al. 1979), since it provides an even sample set that is representative of the sampled space. Its popularity is explained by the fact that a DOE with a desired number of samples can be created, and because of its flexibility (dimensions can be dropped out from the DOE and still have a LHS, because the samples are non-collapsing (Viana 2013). The main drawback of LHS is that it suffers from the curse of dimensionality, where large LHS designs can have inter-variables' correlations (Viana 2013) and space filling properties become questionable. There have appeared some methods to avoid these drawbacks, such as orthogonal arrays and orthogonal LHS, but at the cost of complex optimization algorithms (Viana 2013).

In LHS, for an *N*-dimensional design space, each parameter range is divided in *p* uniformly spaced levels, thus producing $S = p \cdot N$ subspaces. Each level is uniformly sampled only once, ensuring that the full space is sampled, and the resultant number of samples is *p*.

3 Results as Classification and Regression Problems

Whether the purpose of an ABM model is to provide precise quantitative predictions or simply a better understanding of the logical implications of the model hypotheses, the task of analysing the relation between the model output and parameters is usually not simple or easy. The difficulty is greater as the amount of parameters is larger, complicating the use of the traditional graphical techniques to draw inferences. In general, but particularly in models with high dimensional parameter spaces, machine learning techniques can be usefully applied to analyse ABM models.

In order to understand a model, it is absolutely necessary to understand the relationship between the model parameters and the model output. In general, we usually define a statistic representative of the behaviour of the model and analyse the values that it reaches after a number of time steps—the probability function over the set of values—(Izquierdo et al. 2009). In some cases our interest focuses on the asymptotic behaviour of the model for which the statistic is determined by their absorbing states or stochastically stable states (Izquierdo et al. 2009); in others, we want to figure out the state distribution at a time of special interest for the research case study. In any case, the sort of inferences about the statistic, i.e. output variable *Y*, can be described as a function *f* of the model parameters, i.e. input variables $X = (X_1,\ldots, X_p)$ plus an error term of mean zero and independent of *X* (see Eq. 1). This last assumption may be problematic. In such cases, different strategies can be applied depending on the dependence.

$$Y = f(X) + \text{error} \tag{1}$$

We talk about a regression problem if the statistic takes on numerical values (quantitative values); otherwise (qualitative values) we talk about a classification problem. Even with quantitative variables, sometimes the state distribution is composed of a reduced set of states for which the values of the statistic can be grouped in classes, thus turning the problem into a classification problem too. In these cases, classification methods become very useful to explain the relationship between the statistic and the parameters.

Regardless of the type of statistic variable, we can discuss some important issues. The essence of the problem is to estimate the unknown function *f* based on simulation data set. Machine Learning provides a set of parametric and non-parametric methods to solve this supervised learning problem. The selection of a specific learning method usually determines the form of the function $\hat{f}$, used to estimate *f*, so this choice conditions the interpretability of the results. Sometimes *f* takes a linear form, making it easy to understand the influence of the parameters on the statistic; other times *f* is more complex, making the inference more challenging.

The expected test error of a learning method, i.e. the expected error when the estimated function $\hat{f}$ is evaluated on new data not used in the training, can be decomposed in the sum of three terms: the bias error, the variance error and the irreducible error (Hastie et al. 2009). In simple words, the bias error is due to using a function $\hat{f}$ that is not flexible enough to fit the unknown function *f*. The variance error represents the expected change of the estimated function $\hat{f}$ when using different training data. Finally, the irreducible error gathers the natural noise of data, which is the variance of the error term in Eq. (1).

When we seek an estimate $\hat{f}$ of the function *f* we always face a bias-variance trade-off. The goal is choosing a method with the smallest test error, meaning low bias and low variance simultaneously. This issue is related to the flexibility of the learning method, i.e. the degrees of freedom of the function $\hat{f}$. More flexible methods have less bias error, but may overfit data and present higher variance errors than less flexible methods. Frequently the flexibility of a learning method is inversely related with its interpretability, i.e. the ease to explain the relationships between the model output and its parameters. For example, linear regressions are

Table 1 Some popular supervised learning techniques sorted in increasing order of flexibility

Technique
Linear and logistic regression with regularization (Ridge regression, the lasso)
Linear and logistic regression
Kernel smoothing methods
Trees
Boosting methods
Neural networks
Random forests
Support vector machines

more interpretable than kernel smoothing methods (see Table 1), something which is desirable in terms of inference; however, kernel smoothing methods may offer greater flexibility at the expense of overfitting the data losing interpretability.

Assuming that we have selected a function $\hat{f}$ cross validation (CV) techniques (Hastie et al. 2009) can be applied to provide an accurate estimate of the test error. In particular, the k-fold CV divides randomly the data into k subsets of approximately equal size, called folds, and use k-1 of these subsets as training set and the remaining subset as test set. The process is repeated k times, using different subset combination and averaging the results over the test sets (Hastie et al. 2009). If we have several estimating functions to choose from, the same technique can be used to select the best one. Normally, this result is enough to proceed with the inference about the model. However, the estimated test error might not be an unbiased estimate of the performance of the selected function. If an unbiased estimation of the test error of the chosen function is needed, we could apply other refined techniques such as nested CV (Varma and Simon 2006), which develops the essence of CV using two nested loops, an inner loop for function selection and an outer loop for estimating test error.

Table 1 gathers the most important machine learning techniques ordered by increasing flexibility. In general, the election of more flexible methods can be useful when we conduct a preliminary analysis, to get an overall insight to the interactions between all model parameters, while less flexible methods are better for detailed inferences about particular relationships between parameters (Santos et al. 2015). Obviously, the particular research interest always drives the election of the learning method.

4 Variable Importance Analysis

The process of fitting a function f using the parameters as predictors is not only relevant to predict the value of the response variable of interest. Once a regression or a classification model has been adjusted, it is also possible to identify the relevant parameters with the greatest impact on the results. The rationale to conduct this analysis in the context of ABM comes from two related features: (i) the function f can be simplified and hence could be more amenable for interpretation and (ii) the computational effort in subsequent experiments can be focused on those parameters with high influence.

This problem is known in the machine learning and statistics community as feature or variable selection (James et al. 2013). There are several classes of methods to address this problem, e.g., shrinkage or regularization, dimension reduction or subset selection. Some of the subset selection approaches are based on estimating the variable importance for each possible predictor using the fitted function $\hat{f}$. The concept "importance" tries to capture the contribution of each variable to the function f.

In recent years, random forests (Breiman 2001), an ensemble learning method that employs trees as weak learners, have become one of the most popular and widely used techniques in many scientific disciplines. This popularity not only comes from the good predictive performance in classification and regression problems (even in high dimensional and non-linear problems). Random forests are also appealing since they provide a very natural way to find out the importance of each predictor (Criminisi et al. 2011).

A first measure of variable importance in classification problems with ensembles of trees is the decrease in the node impurity measure used to train the trees in each splitting criterion. However a more sophisticated variable importance measure is the "permutation accuracy importance" measure (Strobl et al. 2007). The idea is as follows: given a set of predictor variables X_j used to predict a response variable Y using a random forest, in order to measure the importance of the Xth variable after training, the values of this variable are permuted among the training data and the (out of bag) error is computed and compared before and after the permutation over all trees. This score is sometimes normalized using the standard deviation. Variables with large values are those with higher influence in the response and are therefore considered more important. The underlying assumption of this permutation test is that, by randomly permuting the variable under study, its original association with the response is broken. If that association was originally relevant the prediction accuracy of the forest will decrease, and the higher the decrease the higher the variable importance (Strobl et al. 2007).

Although other variable importance methods are available using different machine learning algorithms (Altmann et al. 2010), the simplicity and interpretability of the permutation test used in random forests and their suitability for being used in complicated situations have made them very popular as a method to gain insight into the significance of the different variables. Notwithstanding, recent research (Strobl et al. 2007, 2008; Wei et al. 2015) has pointed out different sources of bias that can affect this importance measure in random forests. The standard method using CART trees can be hindered in cases where variables vary in their scale level or in their number of categories, or in situations in which there are correlated predictor variables. In those cases, conditional permutation schemes and unbiased conditional inference trees are recommended to reflect the true relevance of the different potential features (Strobl et al. 2007, 2008).

5 Concluding Remarks

The aim of this paper has been to briefly identify some of the concepts from the Machine and Statistical Learning fields that can be useful to the ABM community. We have discussed that this approach can help to improve the understanding of complicated models with many parameters that can be difficult to fully explore. Given that such models can have a huge parameter space and the limited computational resources available, we have discussed some alternatives to explore the parameter space in a more efficient manner.

Considering an ABM model as a function that relates its parameters with its results, one can usefully employ different supervised learning mechanisms to fit such a function. The constructed estimate of the function is an approximation of the agent-based model that can be useful for many purposes. In some cases the approximated fitted function can be more amenable to interpretation and understanding; in other cases it can be useful for generalization and visualization; and very often it can serve to assess the individual impact of the parameters on the results. We have discussed this last analysis process in the case of random forests, but this work is far from being an exhaustive review of all the methods—each one with their advantages and disadvantages—that can be used in the analysis of ABM models.

The insights obtained using these techniques can guide subsequent steps of the modelling process, such as simplifying the model or focusing on a finer grain analysis of the most influential parameters. But they can also be useful from an empirical and a policy-making perspective, leading the efforts of calibration or control into the most relevant variables in the target system.

Acknowledgements The authors would like to thank Dr. L.R. Izquierdo for some advice and comments on this paper. The authors acknowledge support from the Spanish MICINN Project CSD2010-00034 (SimulPast CONSOLIDER-INGENIO 2010) and by the Junta de Castilla y León GREX251-2009.

References

Altmann, A., Tolosi, L., Sander, O., & Lengauer, T. (2010). Permutation importance: A corrected feature importance measure. *Bioinformatics, 26*(10), 1340–1347. doi:10.1093/bioinformatics/btq134

Breiman, L. (2001). Random forests. *Machine Learning, 45*(1), 5–32. doi:10.1023/A:1010933404324

Chang, M. H. (2011). Agent-based modeling and computational experiments in industrial organization: Growing firms and industries in silico. *Eastern Economic Journal, 37*(1), 28.

Criminisi, A., Shotton, J., & Konukoglu, E. (2011). Decision forests: A unified framework for classification, regression, density estimation, manifold learning and semi-supervised learning. *Foundations and Trends in Computer Graphics and Computer Vision, 7*(2–3), 81–227. doi:10.1561/0600000035

Galán, J. M., Izquierdo, L. R., Izquierdo, S. S., Santos, J. I., del Olmo, R., López-Paredes, A., & Edmonds, B. (2009). Errors and artefacts in agent-based modelling. *Journal of Artificial Societies and Social Simulation, 12*(1), 1. http://jasss.soc.surrey.ac.uk/12/1/1.html

Hastie, T., Tibshirani, R., & Friedman, J. H. (2009). *The elements of statistical learning: Data mining, inference, and prediction* (2nd ed.). New York, NY: Springer.

Hernández, C., Galán, J. M., López-Paredes, A., & del Olmo, R. (2014). Economía Artificial: Métodos de inspiración social en la resolución de problemas complejos. *Revista Española de Física, 28*(3), 23–30.

Izquierdo, L. R., Izquierdo, S. S., Galán, J. M., & Santos, J. I. (2009). Techniques to understand computer simulations: Markov chain analysis. *Journal of Artificial Societies and Social Simulation, 12*(1), 6. http://jasss.soc.surrey.ac.uk/12/1/6.html

James, G., Witten, D., Hastie, T., & Tibshirani, R. (2013). *An introduction to statistical learning with applications in R*. New York: Springer.

Lee, J. S., Filatova, T., Ligmann-Zielinska, A., Hassani-Mahmooei, B., Stonedahl, F., Lorscheid, I., Voinov, A., Polhill, G., Sun, Z., & Parker, D. C. (2015). The complexities of agent-based modeling output analysis. *Journal of Artificial Societies and Social Simulation 18*, 4. http://jasss.soc.surrey.ac.uk/18/4/4.html

Macy, M. W., & Willer, R. (2002). From factors to actors: Computational sociology and agent-based modeling. *Annual Review of Sociology, 28*, 143–166.

McKay, M. D., Beckman, R. J., & Conover, W. J. (1979). A comparison of three methods for selecting values of input variables in the analysis of output from a computer code. *Technometrics, 21*(2), 239–245. doi:10.2307/1268522

Santos, J. I., Pereda, M., Zurro, D., Álvarez, M., Caro, J., Galán, J. M., & Briz i Godino, I. (2015). Effect of resource spatial correlation and Hunter-Fisher-Gatherer mobility on social cooperation in Tierra del Fuego. *PLoS ONE, 10*(4), e0121888. doi:10.1371/journal.pone.0121888

Strobl, C., Boulesteix, A. L., Kneib, T., Augustin, T., & Zeileis, A. (2008). Conditional variable importance for random forests. *BMC Bioinformatics, 9*(1), 307. doi:10.1186/1471-2105-9-307

Strobl, C., Boulesteix, A. L., Zeileis, A., & Hothorn, T. (2007). Bias in random forest variable importance measures: Illustrations, sources and a solution. *BMC Bioinformatics, 8*(1), 25. doi:10.1186/1471-2105-8-25

Varma, S., & Simon, R. (2006). Bias in error estimation when using cross-validation for model selection. *BMC Bioinformatics, 7*(1), 91. doi:10.1186/1471-2105-7-91

Viana, F. A. C. (2013). Things you wanted to know about the Latin hypercube design and were afraid to ask. In *10th World Congress on Structural and Multidisciplinary Optimization*, Orlando, Florida, USA.

Wei, P., Lu, Z., & Song, J. (2015). Variable importance analysis: A comprehensive review. *Reliability Engineering and System Safety, 142*, 399–432. doi:10.1016/j.ress.2015.05.018

Chapter 12
Emergence, Culture and Organizational Structure

Marta Posada and Inés Magdaleno

Abstract In this paper, the relationship between structure and culture within an organization is explored using an agent-based approach. In order to analyze how the organizational structure influences the emergence of organizational culture, an agent-based model of bounded-rational agents (who dynamically interact adapting their effort) is proposed. We have found that formal organizational configurations are more likely to favor the emergence of culture than informal interactions.

Keywords Human management resources · Turnover · Culture

1 Introduction

Organizational culture can be defined as "*a system of assumptions, values, norms, and attitudes,... which the members of an organization have developed and adopted through mutual experience and which help them determine the meaning of the world around them and the way they behave in it*". It affects the design of organizational structure (Janićijević 2013).

Organizational structure definitions are as divergent as the approaches for studying them. We follow Dow (1988), who defined organizational structure as "*a relatively stable, either planned or spontaneous, pattern of actions and interactions that organization members undertake for the purpose of achieving the organization's goals*", due to agent-based models consist of artificial agents that interact within an environment (Gilbert 2008).

M. Posada (✉) · I. Magdaleno
Dpto. de Organización de Empresas y CIM, Universidad de Valladolid,
c/Paseo del Cauce 59, 47011 Valladolid, Spain
e-mail: posada@eii.uva.es

I. Magdaleno
e-mail: ines_magdaleno_1989@hotmail.com

C. Hernández (ed.), *Advances in Management Engineering*, Lecture Notes
in Management and Industrial Engineering, DOI 10.1007/978-3-319-55889-9_12

Agent-based modeling is a new analytical method for the social sciences, but one that is quickly becoming popular environment (Gilbert 2008). Although it may bring opportunities and advantages to management and organizations, it is a rather new approach in the management context due to both: (i) variables with complex relationships and complex social interactions are involved, and (ii) achieving a balance between simplicity and realism is not easy (Secchi 2015).

One of the advantages of substituting human subjects by artificial agents, in order to model management and organizations, is that the agents' behavior can be controlled and simplified to the point of being zero-intelligent (see, for an agent-based market context, Posada et al. 2007, 2008). Dal Forno and Merlone (2002, 2004) proposed a multi-agent simulation platform for modeling perfectly rational and bounded rational agents in organizations with an informal organizational structure in order to evaluate the emergence of culture.

In this paper, our aims are to calibrate the effect of different formal organizational structures on the emergence of culture and, by comparing with Dal Forno and Merlone (2002, 2004)'s results, to analyze formal versus informal organization structures.

In the same way as Dal Forno and Merlone (2002, 2004), in our model the employees are characterized by different effort-behaviors and, as a consequence of the interactions between employees, a shared culture of effort can emerges (i.e., an equilibria where all agents exert the same effort). Our model is different from Dal Forno and Merlone (2002, 2004)'s agent-based models in the sense, we introduce a formal organizational structure and explore the relationship between culture and different types of formal organizational structures.

The paper is organized as follows. Section 2 the agent-based model is described. Section 3 the computational experiment settings and results are reported. Concluding remarks are pointed out in Sect. 4.

2 The Agent-Based Model

HURLAB (Human Resource LABoratory) is an agent-based model populated by two types of bounded-rational agents: employees and organization.

2.1 Organization-Agent

The goal of the organization is to achieve the maximum level of average performance with full occupation. Therefore, when an employee leaves the organization it is replaced by other one.

The organization-agent is defined by: *hiring and firing levels* and *organizational structure.*

Hiring—Firing levels: Following Dal Forno and Merlone (2004), both hiring level and firing level are introduced with the objective of achieving the maximum level of performance. The organization dismisses employees whose performance is lower than a firing level.

Organizational structure: The interactions between the employees will be determined by the type of organizational structure. Formal organizational structure ranges from functional to projects depending on the organizations were grouped based on input (functional) or output (projects), respectively, with a variety of matrix structures between them resulting of the combinations of both (Galbraith 2008). We compare functional, balanced matrix and strong matrix (see Fig. 1).

In a functional organizational structure, the employees are grouped into departments and levels, whilst in a matrix organizational structure, the authority is divided into both functional areas and projects.

In matrix structures (which can be weak, balance or strong), there are two roles: the project manager (who coordinates the project) and team members. Team members have to report to multiple project managers (as many as projects each employee is involved) as well as their functional manager. In balanced matrix the project manager has to report to his or her functional manager, while in strong matrix the project manager reports to a multi-project manager.

Besides, in matrix structures, team configuration has to be detailed. Research on team size suggests a curvilinear relationship between team size and team effectiveness. Very small teams (i.e., 2 or 3 people) may lack diversity of perspectives, whereas larger teams (more than 10 members) tend to divide into subteams, which hinders effective team interaction (Dionne et al. 2010). Therefore, we consider a team of 10 people. We consider that a project manager coordinates only one project at a time, and members can participate in 2 or 3 projects. The interactions between the project manager and the members are ego-centered and they are represented by a star-shaped network.

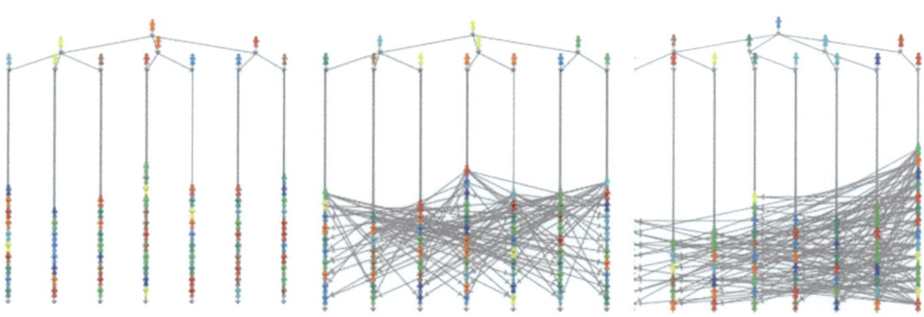

Fig. 1 Organizational structure (*left* to *right*): functional, balanced matrix and strong matrix

2.2 Employee-Agents

Employees' performance is directly estimated by their effort exerted. Employees differ in their effort, which is publicly known. Employees adapt their effort intensity depending on their behavior and as a consequence of their interactions between them.

We consider the following bounded-rational behaviors considered by Dal Forno and Merlone (2002), but the effort ranges from 0 to 100 instead of from 0 to 2: *Null effort*: this agent exerts always the same almost null effort (between 0 and 10); *Shrinking effort*: this agent halves the effort provided by its last partner; *Replicator*: this agent exerts the same effort its last partner exerted in the previous interaction; *High effort*: this agent exerts always the same high effort (between 81 and 100); *Effort comparator:* this agent compares its effort to its previous partner's effort; it increases 10% its effort if it is lower than its partner's effort and vice versa; *Averager*: it averages its effort with last partner's effort.

Employees dynamically interact with the organization and between each other, through a set of behavioral rules. These interactions are defined in terms of the formal organizational configuration. Whom employees interact with is defined by their position in the organizational configuration. Changes in effort are observed by using a graded chromatic scale, which ranges from blue (low effort) to red (high effort).

3 Some Results

The model has been programmed in NetLogo and its description is available at www.eii.uva.es/posada/research.html. In simulations the following parameters are varied: formal configuration structure (functional/balanced matrix/strong matrix), population behavior (homogeneous/heterogeneous), hiring level and firing level. Therefore, many different simulations are possible, but we mention some of the most interesting. The average effort is tracked over time. The changes in the effort values are observed by using a graded chromatic scale, which ranges from blue (for to low effort) to red (for high effort).

We consider an organization populated by 81 employees. The simulation-time ranges from 1 to 10 years. 30 repetitions per combination of parameters are considered to do the statistical analysis.

3.1 Baseline Scenario: Hiring Level = 0, Firing Level = 0

3.1.1 Homogeneous Populations

We have obtained that a homogeneous population consisting of only shrinking effort agents converges to a culture where all the agents play the null effort, independently on the organizational structure, reproducing Dal Forno and Merlone (2002)'s results. We observe in Table 1 that both the average effort in shrinking populations converges to zero and the standard deviation is very low.

We have obtained that a homogeneous population consisting of only *replicator* effort agents, *comparator* effort agents or *averager* agents does not converge to a unique culture, although we observe some groups in *replicator* populations and *averager* population as a consequence of the organizational structure and related to the areas or department. These results are reported in Table 1.

3.1.2 Heterogeneous Populations

Dal Forno and Merlone (2002) found difficulties to observe the emergence of culture in heterogeneous populations. They obtained that the average effort resulting from introducing high effort agents in homogeneous populations depends on the proportion of high effort agents introduced. The influence on null effort homogeneous population is rather predictable (a linear combination of the null effort and the high effort, weighted by the relative proportions of agents). The presence of high effort agents boosts the average effort as can be observed in Fig. 2.

Table 1 Average effort (and standard deviation) in homogenous populations

	Functional	Balanced matrix	Strong matrix
Shrinking	0.0033 (0.0025)	0.0040 (0.0040)	0.0127 (0.0142)
Comparator	36.6891 (5.7972)	36.9780 (5.3299)	36.7365 (6.3739)
Replicator	57.7698 (20.7500)	53.1080 (19.6253)	46.5731 (18.2813)
Averager	52.2322 (10.3311)	48.1795 (12.6582)	48.3803 (9.0852)

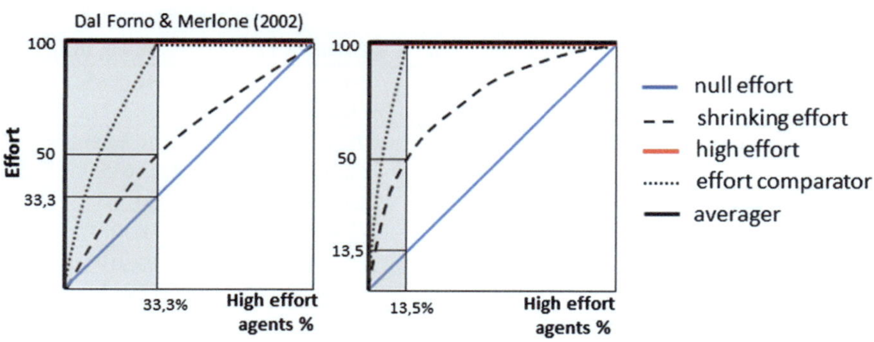

Fig. 2 Supervisor effect

Adapting Dal Forno and Merlone (2002)'s results to efforts range from 0 to 100, the average effort is 33.3 in 66.7% null-33.3% high heterogeneous populations, around 50 in 66.7% shirking-33.3% high heterogeneous populations, and 100 in both 66.7% averager-33.3% high heterogeneous populations and 66.7% comparator-33.3% high heterogeneous populations.

Nevertheless, we have observed emergence of high-effort culture in heterogeneous populations where high effort agents are introduced, in both replicator populations and averager populations, at the supervision level in the organizational structure (see red and orange colors in Fig. 3). The presence of high effort is 13.5%. We observe in Table 2 that the average effort, converges to higher effort than in the homogeneous populations and the standard deviations are very low.

In 13.5% high-86.5% comparator populations a culture does not emerge, (see many colors in Fig. 4) whilst in 13.5% high-86.5% shrinking populations emerge two types of cultures: high-effort culture associated to supervisor level (red color) and other effort culture to the remaining employees (green color) (see Fig. 5).

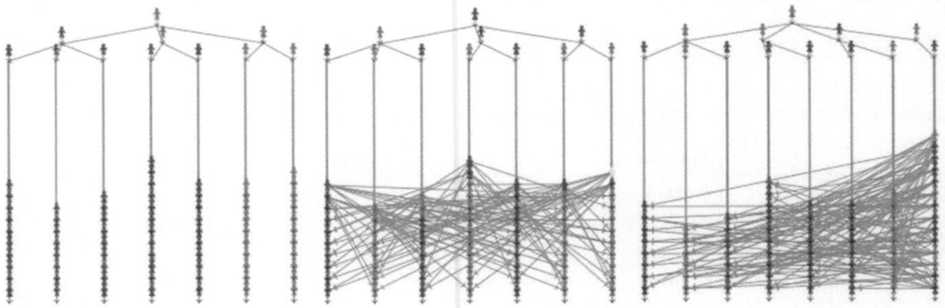

Fig. 3 13.5% of high effort agents (at supervision level) and 86.5% of replicator effort agents

Table 2 Average effort (and standard deviation) in heterogeneous populations

	Functional	Balanced matrix	Strong matrix
13.5% High-86.5% Shrinking	50.9549 (0.8610)	50.1763 (1.2253)	51.0762 (1.2521)
13.5% High-86.5% Comparator	75.8110 (3.3952)	78.5262 (3.3152)	76.8900 (3.6355)
13.5% High-86.5% Replicator	91.1372 (1.1647)	90.4046 (2.4626)	89.2043 (1.5014)
13.5% High-86.5% Averager	90.8238 (0.6586)	88.2619 (2.9515)	89.9196 (2.2134)

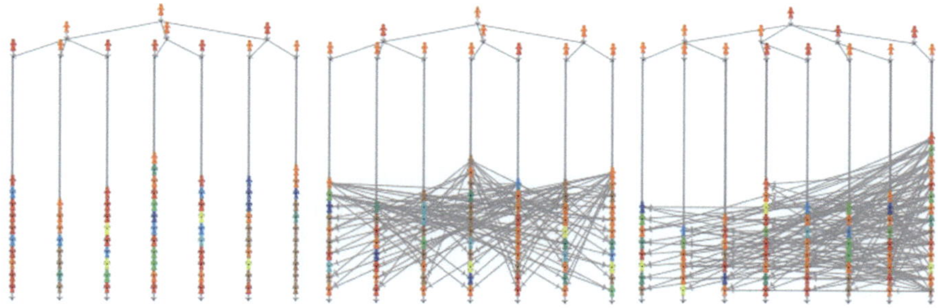

Fig. 4 13.5% of high effort agents (at supervision level) and 86.5% of comparator effort agents

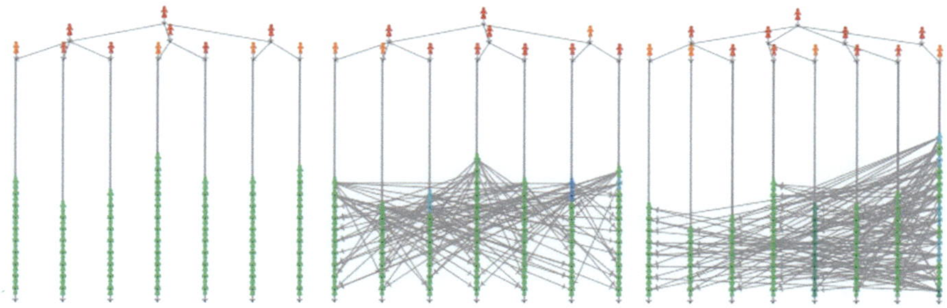

Fig. 5 13.5% of high effort agents (at supervision level) and 86.5% of shirking effort agents

3.2 Firing-Hiring Scenario

In this scenario hiring level and firing levels range from 15 to 50.

When a firing level is considered the emergence of culture may be difficult to observe in both homogeneous and heterogeneous populations, but the average effort increases as it happens in Dal Forno and Merlone (2004).

4 Conclusions

Using a simulation platform, which provides a simple way to study the complex interactions between different types of agents in different types of organizational configurations, we have shown that formal organizational configurations could favor the emergence of culture.

Acknowledgements I would like to thank Prof. Cesáreo Hernández for being my mentor and for having the vision (twenty years ago) to recognize agent-based model as a powerful analytical method to complex context.

References

Dal Forno, A., & Merlone, U. (2002). A multi-agent simulation platform for modeling perfectly rational and bounded-rational agents in organizations. *Journal of Artificial Societies and Social Simulation, 5*(2), 1460–7425.

Dal Forno, A., & Merlone, U. (2004). Personnel turnover in organizations: An agent-based simulation model. *Nonlinear Dynamics, Psychology, and Life Sciences, 8*(2), 205–230.

Dionne, S. D., Sayama, H., Hao, C., & Bush, B. J. (2010). The role of leadership in shared mental model convergence and team performance improvement: An agent-based computational model. *The Leadership Quarterly, 21,* 1035–1049.

Dow, G. K. (1988). Configurational and coactivational views of organizational structure. *Academy of Management Review, 13*(1), 53–64.

Galbraith, J. R. (2008). *Designing matrix organizations that actually work: How IBM, proctor and gamble and others design for success*. Jossey-Bass.

Gilbert, G. N. (2008). *Agent-based models* (No. 153). Sage.

Janićijević, N. (2013). The mutual impact of organizational culture and structure. *Economic Annals, 198,* 35–55.

Posada, M., Hernández, C., & López, A. (2007). An artificial economics view of the Walrasian and Marshallian stability. *Lecture Notes in Economics and Mathematical Systems, 599,* 101–111.

Posada, M., Hernández, C., & López, A. (2008). Testing Marshallian and Walrasian instability with an agent based model. *Advances in Complex Systems, 11,* 249–260.

Secchi, D., & Neumann, M. (2015). A case for agent-based model in organizational behavior. *Team Performance Management, 21*(1/2), 37–50.

Chapter 13
A Study of the Innovative Applications of Intelligent Transport Systems Works to Logistics and Freight Transport: Public-Private Collaboration Projects

Moisés Javato-Martín, Pedro Sanz-Angulo, Juan José de Benito-Martín and Jesús Galindo-Melero

Abstract Intelligent Transport Systems are fundamental to enterprise competitiveness, especially in terms of the efficiency improvement of Logistics and Freight Transport. Although these areas are commonly associated with the private sector, the public sector has a significant role regarding regulation, and the planning and maintenance of public transport infrastructures. In this paper, two innovative areas in which public and private sectors collaborate are described: namely, gateway facilitation technologies, and some projects focused on an innovative management of freight transport.

Keywords Intelligent transport systems · Logistics · Freight transport · Public-Private collaboration

1 Introduction

The fact that transport is an economic sector which makes a significant contribution to the improvement of the competitiveness and the efficiency of an economy is something unquestionable today. The European Union may serve as an example.

M. Javato-Martín · P. Sanz-Angulo (✉) · J.J. de Benito-Martín · J. Galindo-Melero
Department of Business Organization and Marketing and Market Research,
School of Industrial Engineering, University of Valladolid, Paseo del Cauce,
59, 47011 Valladolid, Spain
e-mail: psangulo@eii.uva.es

M. Javato-Martín
e-mail: moisesjavato@yahoo.es

J.J. de Benito-Martín
e-mail: debenito@eii.uva.es

J. Galindo-Melero
e-mail: jesus.galindo@uva.es

C. Hernández (ed.), *Advances in Management Engineering*, Lecture Notes in Management and Industrial Engineering, DOI 10.1007/978-3-319-55889-9_13

Thus, in the EU the transport industry directly employs more than 10 million people, accounting for 4.5% of total employment, and represents 4.6% of Gross Domestic Product (GDP); the manufacture of transport equipment provides an additional 1.7% GDP and 1.5% employment (European Commission 2011a). Nevertheless, although transport is indispensable and beneficial to society, it is also the cause of several problems, the most important of which are traffic congestion, pollution and road accidents. The use of technology may reduce the effects of those main problems and, at the same time, may provide new competitive solutions to companies.

Therefore, this is the context in which Intelligent Transport Systems (ITS) have appeared. They may be defined as “a series of advanced applications derived from computing, electronics, and communications technologies which, from a social, economic, and environmental point of view, are intended to improve transport mobility, safety, and productivity, optimizing the use of existing infrastructures, increasing energy consumption efficiency and improving the capacity of transport systems” (Ortiz 2010, p. 14). Although they can be applied to any mode of transport (ground, ship or air), in practice its development is mainly associated to road transport, which is the area covered in this paper.

A key element in Intelligent Transport Systems is information and its management. Without a trace of a doubt, all players of the logistics sector (the shipper, the carrier, and the consignee) are interested in having quality information (i.e., precise, complete and on time) regarding shipments and transport operations. Accordingly, the old saying that “the information is as important as the cargo” is still valid (Wolfe and Troup 2013). ITS provide ways of obtaining and using this information, generating benefits not only for logistics companies but also for society as a whole.

Although freight transport is a private business, the public sector has a significant role regarding regulation, and the planning and maintenance of public transport infrastructures. In contrast with the private sector, which is mainly interested in finding more efficient and reliable ways of delivering the cargo, the public sector is more oriented towards applications related to safety assurance and vehicle electronic screening. Thus, while in the private sector the most typical application is fleet management (i.e., the tracking of trucks and containers), in the public sector the leading systems are inspection technologies (e.g., electronic weighing and credentials checking).

The study of fleet management applications is not considered in this paper as these applications are currently well developed.[1] Instead, two innovative areas in which public and private sectors collaborate are described: gateway facilitation technologies, and the projects for an innovative management of freight transport.

---

[1]An example of those applications is the MOVILOC® system of the company GMV, which allows real-time positioning and tracking of vehicle fleets (www.moviloc.com).

2 Gateway Facilitation Technologies

In the United States, pioneering ITS applications which facilitate the inspection and access control of commercial vehicles have been conceived. They are known as Gateway Facilitation Technologies and are based on a programme of the US Department of Transportation (USDOT) called CVISN (Commercial Vehicle Information System and Networks). This programme defines an architecture through which centre of the public administration (Commercial Vehicle Administration Centers) and those of private companies (Fleet and Freight Management Centers) are able to exchange information related to credentials, taxes, and drivers. A series of services which will use the information are also defined in the programme.

There are two kinds of services within the CVISN programme: core CVISN capabilities, which are well established and defined, and expanded CVISN capabilities, to be determined by the company which provides the services, as long as they comply with the standards of the CVSIN architecture (Wolfe and Troup 2013, p. 26). Core CVISN capabilities are basically three: safety information exchange, credential administration and electronic screening. Some examples of expanded CVISN capabilities are: virtual weigh stations (i.e., facilities that do not require continuous staffing and are monitored from another location), license plate readers or one-stop shops (i.e., web portals where users log in once to access several public systems). An expanded application which is also worth mentioning is weigh-in-motion, which is related to the experimental programme Smart Road Initiative (Hill 2013).

Currently, there are two main commercial applications based on the CVISN programme: PrePass and PierPass. PrePass (www.prepass.com) is an American Automatic Vehicle Identification (AVI) system for the inspection of the trucks equipped with a wireless transponder, as they approach weigh stations or other inspection sites. If the result of the wireless checking is successful, the vehicle could continue its course without speeding down or entering the inspection site. Consequently, the truck traffic through the site decreases and station resources and staff can be used more efficiently, paying more attention to the vehicles which really need it. The vehicles which use Prepass need to be certified in advance: its credentials and safety records must be periodically verified by the authorities, in order to validate that they meet the terms of state and federal regulations.

The PrePass programme is funded and managed by a public-private partnership, HELP Inc., which comprises the US Department of Transportation (USDOT), several states, and private companies.[2] The specific operation of the system is shown in Fig. 1. First, when a vehicle approaches the inspection site, an electronic reader on a boom over the motorway automatically reads the transponder information and identifies the vehicle (1). Then, a computer in the inspection site

[2]Presently, PrePass includes 301 inspection and weigh station in 31 states. More information about the programme can be found in (PrePass Marketing 2009).

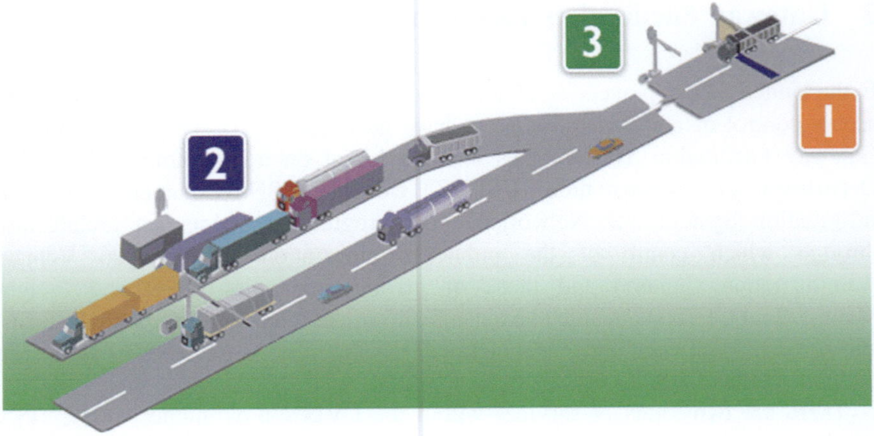

Fig. 1 The operation of the PrePass system (PrePass n.d.)

retrieves the information associated with the transponder, and validates it in order to check that the vehicle comply with the state regulations (2); sometimes a weigh-in-motion inspection is also performed in order to verify the weight of the truck. Finally, when the vehicle reaches a second electronic unit (3) a signal is transmitted to the vehicle transponder in order to tell the driver what to do: keep going (if a green light is displayed) or enter the inspection site (if the transponder shows a red light along with a beep).

In the field of port access control, there is a nongovernmental initiative based on CVISN which should be taken into account: PierPASS (www.pierpass.org/). It is a not-for-profit organization created by the Marine Terminal Operators (MTO) in the ports of Los Angeles and Long Beach. Both ports are the highest volume intermodal container ports in the US, and are located in an area which is infamous for traffic congestion (Metropolitan Los Angeles); therefore, finding ways to facilitate the access of trucks to the port is essential for mitigating the ports' contributions to congestion. In this regard, PierPass addresses port operating efficiencies, road and highway congestion, air quality, and port security concerns (Wolfe and Troup 2013).

The PierPass system is based on a RFID tag attached to the truck (Truck Tag) in order to improve port security and to facilitate the operations at the terminal gate. This tag is similar to those used for toll collection (e.g., E-ZPass) and is mounted on the wing mirror of the truck. In order to use the PierPass systems the trucks must pass an inspection programme (Truck Check) which is performed by the external company eModal. When the truck arrives at the maritime terminal gate, the truck tag is read in order to verify that the truck and its driver have the security clearance to go into the terminal. This process is carried out automatically without the intervention of port staff, so the waiting time at the gate decreases substantially.

3 Projects for an Innovative Management of Freight Transport

The first main research and development project to be considered is e-Freight (European e-Freight Capabilities for Co-modal Transport) (www.efreightproject.eu), which is co-financed by the European Union under the 7th Framework Programme. Starting the 1st of January 2010, its work plan has covered 4 years and comprises 31 partners from transport-related industries and includes the involvement of 14 EU member States and Norway. This project aims to develop one of the initiatives which were included in the White Paper adopted by the European Commission in 2011, in particular the initiative number 7: "Create the appropriate framework to allow tracing goods in real time, ensure intermodal liability and promote clean freight transport" (European Commission 2011b).

In accordance with such objective, the project considers the development of the following elements: a standard framework for freight information exchange covering all transport modes and all stakeholders (e-Freight Framework); a single European transport document for all carriage of goods, irrespective of mode should be developed along with all the necessary legislative support; and a single window (single access point) and one stop shopping for administrative procedures in all modes. The underlying vision of the e-Freight project is a future Europe where transport and logistics operations should be handled electronically, in a paper-free manner, and where all transport modes should connect safely and seamlessly.

A practical application of this project is described in (eBos 2013). In the latter, the Danish transport and logistics company DSV and the Swedish ferry operator Stena Line combine the e-Freight framework and the standards developed by the organization GS1 (www.gs1.org) in order to improve the efficiency of their operations. This combination is illustrated by means of an example: the transport company DSV receives an order from one of its clients, the Swedish paper products manufacturer SCA, which requests a delivery to Germany. In order to complete such delivery, DSV has to manage multimodal transport (both road and ferry), and the information generated in the process is shared by all stakeholders using a centralized software platform which is based on the e-Freight framework.

Once the order from the client is received, DSV sends a standard GS1 message (Transport Instruction Message) to Stena Line in order to book enough capacity in one of its Germany-bound ferries. After receiving the booking confirmation, DSV chooses the freight company which will convey the cargo to the port, and sends a message to that company including the booking confirmation number provided by Stena Line. This number, as well as the truck driver information and the estimated time of arrival to the port, is sent to Stena Line by the freight company using another message, so that the ferry company should be prepared.

At the same time, the cargo to be sent is processed in the warehouses of the client who requested the transport. The goods are labelled, scanned using GS1-format tags (triggering, an automatic message to DSV) and loaded into the truck. Reading the tags is the first step of a process by which the arrival of the shipment is confirmed, by means of messages, in each stage of the route. When the loaded truck arrives to the port, its driver goes into the terminal directly by introducing the booking number previously received, and delivers the trailer. After completing the security check, the trailer is shipped into the Stena Line ferry, and a confirmation message is sent to DSV. Border crossing is notified to the foreign authorities using a CRS message (Common Reporting Schema) which was developed as part of the single window component of the e-Freight project. As soon as the trailer arrives in Germany, it is picked up by the local freight carrier, and a status message is sent to DSV. Finally, when the cargo reaches its final destination, a German supermarket, DSV receives the last confirmation message.

The second main research and development project to be considered is iCargo (Intelligent Cargo in Efficient and Sustainable Global Logistics). It is a four-year collaborative project (2012–2015) funded by the European Union under the 7th Framework Programme and comprises twenty nine organizations with expertise in logistics and ICT, including commercial operators, trade associations, research organizations and public authorities. The iCargo project aims to support the evolution of the logistics industry towards a mature business ecosystem, based on cooperation between specialized actors to offer competitive and efficient door-to-door logistics solutions (ECITL 2014).

The project is intended "to build an open affordable ICT infrastructure that allows real world objects, existing systems and new applications to efficiently cooperate, enabling mores cost effective and low CO_2 logistics through improved synchronization and load factors across all transport modes" (iCargo, n.d.). The objective of this architecture is to support new logistics services that synchronize vehicle movements and logistics operations across various modes and actors to lower CO_2 emissions; adapt to changing conditions through dynamic planning methods involving intelligent cargo, vehicle and infrastructure systems; and combine services, resources and information from different stakeholders, taking part in an open freight management ecosystem.

As part of the ecosystem which iCargo wishes to create there is the notion of "Intelligent Cargo": transforming goods into intelligent cargo items from which real-time information can be obtained in order to actively manage transport and to improve performance. This idea is already present in the e-Freight Project, which envisages a future where goods should become self-context and location-aware (e.g., aware of traffic and infrastructure conditions) and should be connected to a wide range of information services, in an automatic transport management process (www.efreightproject.eu).

One of the central purposes of the project is to create a core reference model, the Common Framework, which allows ecosystem members to interact by mediating between the different standards and systems in use. In addition to research and technological development, iCargo has included the implementation, demonstration and validation of real pilot projects. Two of the most important pilots are explained here: the DHL Business Case and the Scandinavia-Continental Europe pilot project.

3.1 The DHL Business Case

The logistics provider DHL has implemented a load optimization system for road freight transport, which is based on the use of a software platform and cross-docking centres of several logistics and manufacturing companies all over Europe.

It is estimated that 20% of the trucks running in Europe are empty, with a result of an average load factor close to 50%. This fact has significant environmental and economic consequences; thus, the potential benefits of optimizing transport would be an abatement of 124 mega-tons of CO_2 per year and cost savings from 100 to 300 billion euros per year (iCargo Project 2015).

As described in Voorspuij (2013), the logistics provider DHL along with other iCargo partners has undertaken the task of addressing this and others logistic waste problems (Table 1) without introducing significant changes or investments in their companies.

The solution adopted is based on a software platform, known as MixMoveMacht (https://www.mixmovematch.com/), which was conceived by the company 3 M and developed by the Swedish consulting firm Marlo, following the GS1 Standards and the iCargo Framework. It is an open platform, designed to be integrated with the software and IT products of any partner involved (e.g., with their ERP).

The project is based on a cargo consolidation-deconsolidation technique (Fig. 2) which is implemented in cross-docking centres called Reconstruction Hubs and is optimized by the MixMoveMatch software.

The operation of the system is as follows. Each manufacturer or client conveys the goods to be shipped to the logistics provider in individual packages (i.e., small pallets or boxes). The logistics provider labels each individual package with a unique barcode (SSCC), and this code is scanned in order to check that the label information matches the EDI data sent by the client, avoiding shipment errors from the beginning.

The packages from different clients are grouped together to form complete big pallets; this is the first element of the process (Mix). Then the pallets are transported following a route which may include one or more reconstruction centres until they

Table 1 Logistics waste problems addressed by DHL (Voorspuij 2013)

Problem	Objective
Wasting fuel	Fewer trucks on the road for the same cargo volume
Relabeling pallets and cartons by parties involved in end-to-end movement of goods	Using of a single end-to-end label attached by the shipper
Rekeying of information already captured elsewhere	Using a common standard software platform
Wasting time correcting errors and solving confusion	Using an open, independent, comprehensive system of standards
Wasting transportation capacity due to the lack of coordination among players in the supply chain	Creating a trustworthy environment for competitors to collaborate

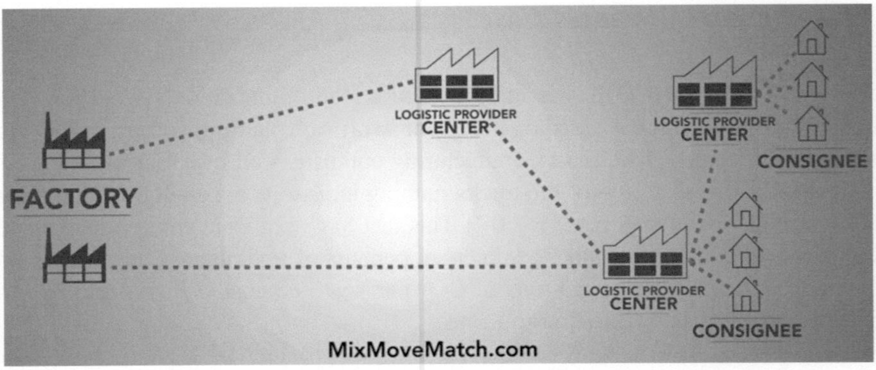

Fig. 2 The MixMoveMatch scheme (iCargo Project 2015b)

get close to their final destination; this is the second element (Move). In the last centres, the individual packages are extracted from the big pallets and classified in order to "reconstruct" the specific shipments of each client; this is the last element (Match). As a result of this process, the transport is completed using more efficiently loaded trucks.

It should be noted that the reconstruction hubs which have been implemented all along Europe do not belong exclusively to DHL, but also to other companies such as 3 M, Ceva, DPD or Spedition Kleine. The latter demonstrates that logistics providers and shippers are able to share resources and optimize their operations for their mutual benefit, using a common software platform and standard labelling.

Regarding the results of the project, the company 3 M estimates a reduction of its transport costs by 35% and of its CO_2 emissions by 50% since the system was launched. Likewise, DHL estimates a warehouse dispatch area reduction of 50% and an inbound processing time reduction up to 50% (iCargo Project 2015b).

3.2 *The Scandinavia-Continental Europe Pilot Project*

The second real project which comes from the application of iCargo is a pilot programme deployed in Sweden by the Swedish Transport Administration (Trafikverket), the paper product company Stora Enso and the American consulting firm IntelliTrans. The goal of the programme is to put the iCargo concepts into practice in order to improve the efficiency of the deliveries which Stora Enso makes from Sweden to the rest of Europe by rail and ship.

The project profits from an existing RFID system already installed in the Sweden railways, called RFID in Rail (Ivansson 2012). This system uses RFID readers which are mounted alongside the tracks and RFID tags which are attached to the wagons (Fig. 3) in order to follow the movement of the cargo. The information

Fig. 3 RFID reader and tag attached to a wagon of the Swedish railways (Ivansson 2012)

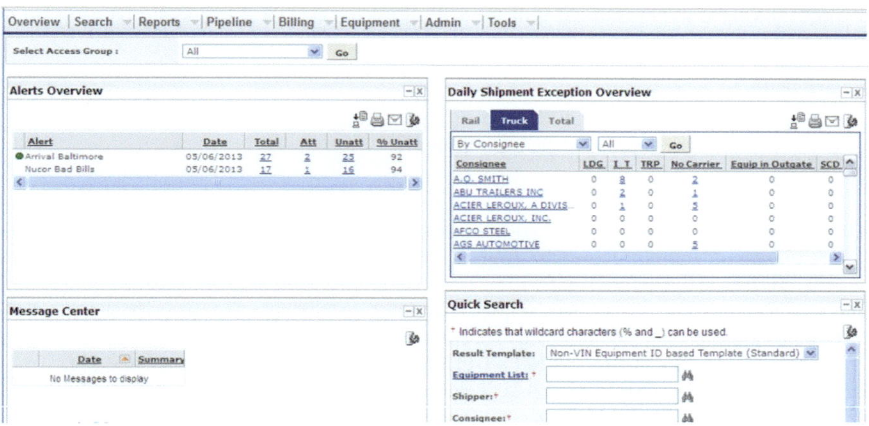

Fig. 4 Alert window from the software of the iCargo Swedish pilot project (IntelliTrans 2014)

obtained in the progress is structured and transmitted using GS1 Standards. The operation of the system is explained in (iCargo Project 2015).

An intelligent corridor based on this RFID system has been established by the firm IntelliTrans in collaboration with Stora Enso. It allows Stora Enso to track the wagons carrying its products as they travel from its factory to the Gothenburg port, from which the goods are sent out of Sweden.

The consulting firm IntelliTrans has developed a software platform, called GVP (Global Visibility Platform), which compares the wagon information detected by the RFID readers with the railway schedule and alerts stakeholders (e.g., clients, suppliers, carriers and so on) in case of delay (Fig. 4).

The platform is also capable of dynamically rescheduling the bookings for the cargo with the shipping company at the port, when a railway delay is detected. This contributes to a more effective use of the freight capacity of the ships (IntelliTrans 2014).

4 Conclusions

We have described and commented two innovative areas in which private and public sectors are collaborating: gateway facilitation technologies and outstanding projects related with innovative management of freight transport. We may conclude that:

The use and development of ITS can generate significant benefits in terms of transport efficiency, customer satisfaction, and environmental sustainability.

Several projects at European level have demonstrated through applications and concrete examples, the efficiency of these systems, providing significant advantages.

The projects described here, although with different approaches, maintain a fairly similar objectives, and use the same GS1 standard messages.

In summary, because of the potential benefits to be gained from the implementation of ITS, it would be desirable the involvement of public institutions to promote a single ITS standard that would allow the use of these systems as soon as possible.

References

ECITL (2014). ICT drives transport logistics cooperation. In *Conference book of the 7th European Conference on ICT for Transport Logistics*. Dortmund. Retrieved from http://www.ecitl.eu/ecitl_2014/ECITL_confbook.pdf

European Commission (2011). White paper roadmap to a single European transport area—Towards a competitive and resource efficient transport system. Brussels. Retrieved from http://eur-lex.europa.eu/legal-content/EN/TXT/PDF/?uri=CELEX:52011SC0391&from=EN

European Commission (2011, April 14). Transport matters. Retrieved from http://ec.europa.eu/transport/strategies/facts-and-figures/transport-matters/index_en.htm

eBos (2013, March 20). GS1—e-Freight. Optimizing intermodal transport by combining GS1 Standards and the e-Freight Framework [Video]. Retrieved from https://vimeo.com/62234636

Hill, C. (2013). Module 13: Connected Vehicles. ePrimer—ITS Professional Capacity Building Program. United States Department of Transportation (USDOT). Retrieved from http://www.pcb.its.dot.gov/eprimer/module13.aspx

iCargo Project (2015, February 9). RFID in Rail—On the right track with RFID [Video]. Retrieved from https://youtu.be/TF2d-_kzRg4

iCargo Project (2015b, March 23). MixMoveMatch 23 Mar HD [Video]. Retrieved from https://youtu.be/nxAkfTsAM9k

iCargo (n.d.). About iCargo Project. Retrieved June 14, 2015, from http://i-cargo.eu/content/about-icargo-project

IntelliTrans (2014, December). iCargo in action. Retrieved from http://www.intellitrans.se/uploads/iCargo.pdf

Ivansson, G. (2012, October 31). RFID in Rail. RFID Journal. Retrieved from http://www.rfidjournal.net/masterPresentations/rfid_europe2012_oslo/np/ivansson_315_oct24_gs.pdf

Ortiz, M.C. (2010). Los Sistemas Inteligentes de Transporte. Su aplicación a los modos terrestre, marítimo y aéreo. Madrid: Secretaría General de Transporte. Ministerio de Fomento.

PierPass (2014, April 13). Gate to Gate: What Happens When a Truck Picks Up a Container? Retrieved from https://youtu.be/P9IJN1yIIJ4

PrePass Marketing. (2009, June 30). PrePass Private Sector [Video]. Retrieved from https://youtu.be/DugHQfjkaEY

PrePass (n.d.). What Is PrePass. Retrieved June 14, 2015, from http://www.prepass.com/services/prepass/Pages/WhatIsPrepass.aspx

Voorspuij, J. (2013, June 4). Eliminating waste in logistics. Mix Move Match—A good practice to address structural waste in many supply chains. Retrieved from http://i-cargo.eu/content/eliminating-waste-logistics

Wolfe, M., & Troup, K. F. (2013). Module 6: Freight, Intermodal, and Commercial Vehicle Operations. ePrimer—ITS Professional Capacity Building Program. United States Department of Transportation (USDOT). Retrieved from http://www.pcb.its.dot.gov/eprimer/module6.aspx

Part IV
Innovative Teaching Activities in Management Engineering

Chapter 14
Lean School: A Practical Space of Cooperative Learning from the Factory to the University

Ángel Manuel Gento, Juan José de Benito, Pedro Sanz and José Antonio Pascual

Abstract In recent years, developing experiential learning has fulfilled the requirement that engineering students fully understand the concept of Lean Manufacturing, or Lean Production, by demonstrating the advantages and disadvantages of some of their key principles. Learning Factories have been developed to provide students and industrial participants with hands on instruction to learn a manufacturing system that produces small-scale models. In our paper, we describe the Lean School developed in conjunction with an industry partner (Renault) to improve the capabilities of our College of Engineering students and of workers in companies located in the Castile-León region.

Keywords Lean manufacturing · Learning factory · Learning-by-doing · Laboratory

1 Introduction

Kiichiro Toyoda, Toyota's president and founder, realised the urgent need to increase productivity in order to ensure the company's survival. To achieve its mission Toyota began to place a greater emphasis on identifying and eliminating sources of waste within the company. Like any cultural change, this process is

Á.M. Gento (✉) · J.J. de Benito · P. Sanz · J.A. Pascual
Dpto. de Organización de Empresas Y CIM, Escuela de Ingenieros Industriales, Universidad de Valladolid, Paseo del Cauce, 59, 47011 Valladolid, Spain
e-mail: gento@eii.uva.es

J.J. de Benito
e-mail: debenito@eii.uva.es

P. Sanz
e-mail: psangulo@eii.uva.es

J.A. Pascual
e-mail: pascual@eii.uva.es

C. Hernández (ed.), *Advances in Management Engineering*, Lecture Notes in Management and Industrial Engineering, DOI 10.1007/978-3-319-55889-9_14

based on people, respect for people, and respect for their experiences and opinions. Anticipating that the innovations they were introducing would most likely have significant cultural impact, as had been the case even in the company's infancy, Toyota leadership made a commitment to base their new production process on respect for the values, experiences, and opinions of the people involved in and affected by their proposed changes in the current manufacturing status quo.

In western culture, this philosophy gained recognition during the oil crisis of the 70s. During this period, American car factories noticed that this critical world situation had less of an impact on the Japanese counterpart. In 1990 Womack, Jones, and Roos of the Massachusetts Institute of Technology studied the concepts and methodologies of the Toyota Production Systems, and adapted them to the western culture. Thus was born the term Lean Manufacturing (LM).

LM has been so successful in the automotive industry that it has become a benchmark not only in production but in service provision as well. Other fields have customised the terms and processes to fit their respective disciplines. For example, see Lean Service (Bowen and Youngdahl 1998), Lean Healthcare (Bushell and Shelest 2002), Lean Logistics (Jones et al. 1997) or Lean Accounting (Hines et al. 2002).

A study by Lamancusa et al. (2008) verified that Lean Management (LM) concepts have been implemented efficiently, though theoretically, in the university environment. Therefore, the University of Valladolid Department of Management, in conjunction with Renault's consulting company, have created a space to implement the principles of operational excellence (Lean System) in a small scale business environment similar to those already developed in other countries and universities (De Zan et al. 2015; Matt et al. 2014; Oberhausen and Plapper 2015; Pozzi et al. 2015; Veza et al. 2015).

"The potential of learning factories to develop softer competencies, for instance in the field of change management, has been identified by Wagner et al." (Dinkelmann et al. 2014). Learning factories offer "learning by doing" possibilities, so it is possible to test the best practices and the worst situations on realistic models that adequately illustrate the possibilities and limitations of the Lean tools.

2 Lean Manufacturing

There are many definitions of LM. We can consider that Lean is a philosophy that facilitates continuous improvement (kaizen) and removal of waste in a manufacturing system (Ohno 1988). Lean difference three types of waste:

- Muda: any activity in a process that does not add value.
- Mura: any activity created through unbalanced situations.
- Muri: any activity asking unreasonable stress or effort from personnel, material or equipment.

The seven most common types of waste are transport, time/waiting, motion, inventory, over-processing, over-production and defects (Fig. 1) (Womack et al.

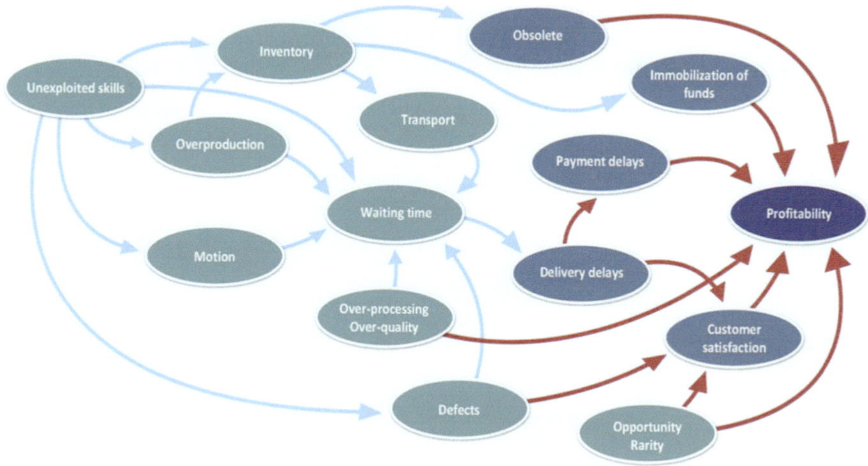

Fig. 1 Map of interdependence of 7 wastes (Klym 2013)

1990). Actually, we can consider 7 + 1 and 7 + 2 wastes to be, respectively, unexploited skills and change resistance of managers. If we can eliminate or reduce these wastes we improve overall customer value.

Typically, value-added activities change the form, fit, or function of a product or service. These are things for which the customer is willing to pay. Activities that do not add value to the processes are called non-value-added activities and can be eliminated, reduced, or simplified. In many cases, the non-value activities account for 95% of total lead-time (Fig. 2).

Lean approach involves 5 stages:

1. Identify the customer values regarding products' and modify processes to reflect those values.
2. Integrate the value stream by identifying value added and non-value-added activities.
3. Remove waste, improve product flow, and optimise batch sizes.
4. Adjust production to consumer demand.
5. Maintain self-sustaining continuous process improvement.

To apply Lean School principles, students must learn the most common lean tools, as proposed by Helleno et al. (2013): 5S's, Bottleneck Analysis, Continuous Flow, Gemba, Heijunka, Hoshin Kanri (Policy Deployment), Jidoka, Just In Time (JIT), Kanban (Pull System), Kaizen (Continuous Improvement), Root Cause Analysis (Ishikawa diagram), Poka-Yoke, Single Minute Exchange Die (SMED), Value Stream Map (VSM), Plan-Do-Check-Act (PDCA), Visual Control, Spaghetti charts, Failure Mode, and Effects Analysis (FMEA).

To evaluate the performance of the implemented processes, the most important indicators or key performance index (KPI's) are calculated: Overall Equipment Effectiveness (OEE), Takt Time, Cycle Time, Total Effective Equipment

ACTIVITY		INDUSTRIAL PROCESS	CATERING SERVICES	MEDICAL SERVICES	ADMINISTRATIVE PROCESS
Operation	○	Connection of the motor in a washing machine	Addition of sauce to a fish dish	Radiography	Introduction of the information of an invoice in the computer
Inspect	□	Quality control of the connection	Control of flavor and PH of the sauce	Control of patient identity data	Checking the invoice data
Delay	D	Working places waiting for the supply of cables	Fish waiting for the addition of the sauce	Patient in a stretcher waiting for the doctor	Invoice waiting for the signature of the responsible person
Inventory / Warehouse	▽	Washing machines on AGV or in warehouse	Bowl with elaborated sauce for several fish dishes	Waiting area with patients waiting to be called	Folder with invoices to introducing in the computer
Move	⇨	AGV with washing machines to packing area	Fish dishes with sauce taken to the customer´s table	Stretcher with patient moved to an operating room	Folder of invoices taken to filing cabinet

Activities that add no value

Fig. 2 Symbols defined by ASME (American Society of Mechanical Engineers) before 1957 and adopted in BSI: Glossary of terms used in management services, BSI 3138 (ILO 1992)

Performance (TEEP), Direct and Indirect Labour, Availability, Performance, Quality, Overall Labor Effectiveness (OLE), …

3 Valladolid University

The University of Valladolid (UVA) (www.uva.es) is a public university in the city of Valladolid, province of Valladolid, in the autonomous region of Castile-Leon, Spain. Established in the 13th century, it is one of the oldest universities in the world. The university has 32,000 undergraduate students and more than 2,000 teachers.

Today UVA is one of the leading centers of Higher Education in Spain, offering over 100 degrees, 17 post-graduate programs, and 50 masters degrees, and boasting a broad network of international relations, research centres, wide-ranging sports, cultural facilities, and a rich architectural and documentary heritage. All of these factors combine to provide an outstanding academic environment, equal to the oldest universities in Europe, in terms of history, excellence, and quality of research.

In Europe, the Bologna Accords in 2005/2006 gave a new structure to the higher education system, proposing two education cycles: undergraduate and graduate. The undergraduate study cycle lasts three (or four) years (Bachelor) and the graduate cycle is two years (or one year) of specialization, in which the student obtains the graduate certificate (Master). This new structure applied to engineering education results in, according to Salvador and Oliveira (2012), a time reduction but also deficiencies in skills that are necessary for graduates to adequately meet industry needs.

The College of Engineering (www.eii.uva.es) is one of the largest at UVA. To improve the practical training of students in a realistic production environment (International Monetary Fund 2013), we have developed, in collaboration with Renault's consulting company (www.reanult-consulting.es) the Lean School.

According to Dale's Cone of Experience (1969), the most effective learning process is the one that is able to provide as many as possible links to practical and concrete experiences. It is worth reporting that people generally remember only 10% of what they read, 20% of what they hear, 30% of what they see, 50% of what they see and hear, 70% of what they say and write and over 90% of what they do.

The Lean School can be used for educating academic students as well as to train industrial participants, thereby improving knowledge transfer between universities and industry. The objective of the Lean School is to offer participants a learning environment that allows a balanced relationship in conveying specialized theoretical and analytical knowledge as well as hands-on experience (Seitz and Nyhuis 2015).

4 Lean School

The formation in the Lean School is especially focused in the practice of the Lean tools adapted to the managerial environment of the UVA. Therefore, products are available both in the industrial sector (cars, streetlights) and the food industry (pasta, cookies, and jellies).

Lean School consists of a workshop area composed of the following areas: machining, stamping, and assembly (Fig. 3). Each of these areas contain different industrial and educational resources necessary for the manufacture and assembly of the product: jobs and tools, handling and transport of equipment, storage and packaging equipment, etc., and consumables for the adaptation of equipment to each specific situation by students.

The production area is made up of sections, similar to departments in a manufacturing plant. Figure 3 illustrates a typical workplace configuration:

- Machining Department (red-continuous rectangle): 1 stamping workplace and 4 machining workplaces.
- Assembly Department (blue-dashed rectangle): with 3 areas and workplaces.
- Three storage areas (yellow-dotted rectangles): raw material, in progress materials, and finished materials.

The modular configuration of the different positions allows students to change the configuration quickly from one to another (based on his progress). Actually, we can define four phases in the Lean School to observe and correct the 7 + 1 typical wastes of Lean:

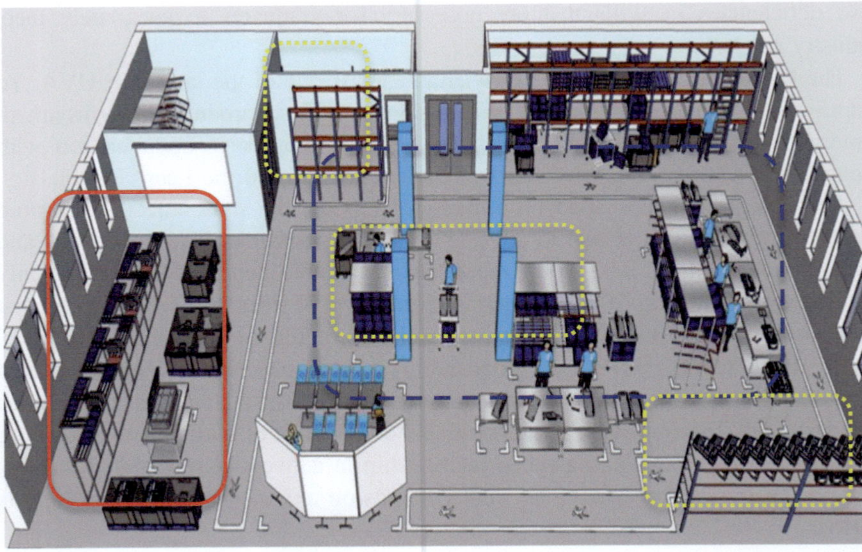

Fig. 3 3D image of lean school

1. Flow *PUSH* disorganized, with low standardization (Fig. 3).
2. Flow *PUSH* organized, where the students ensure the quality of production and reassign activities to balance workloads.
3. Flow *PULL* unoptimized, working with the one-piece flow.
4. Flow *PULL* optimized, and the one-piece flow (Fig. 4).

The laboratory currently offers two products for student practice: cars (Fig. 5) or streetlights (Fig. 6). Students are assigned to a product depending on the individual student's skill level and the allowed time for practice.

Students receive a manual that contains a list of parts needed to manufacture different products, the Operation Process Charts (Fig. 7), and the quality controls required by customers. The student must develop the Flowchart and Flow Process Chart diagrams.

After a brief training session in assembly, students form six groups of two people each to work in different sites. An additional group of two-three people are in charge of logistics for the different sites and the last group of two are responsible for deciding different work site activities and the production order.[1]

[1]This kind of manufacturing process is not atypical. For instance, companies with production technologies (e.g., cutting, machining, painting, testing), typically are grouped into departments, whereby products travel in batches through many stages. In addition, production processes usually involve different technologies and several companies. Even, there are many examples of companies that have their departments (e.g., sales, purchasing, R&D, HR, finance) distributed geographically as well as many suppliers and/or partners with a lasting unreliable delivery time.

Fig. 4 Lean school final manufacturing line (flow "pull" optimized)

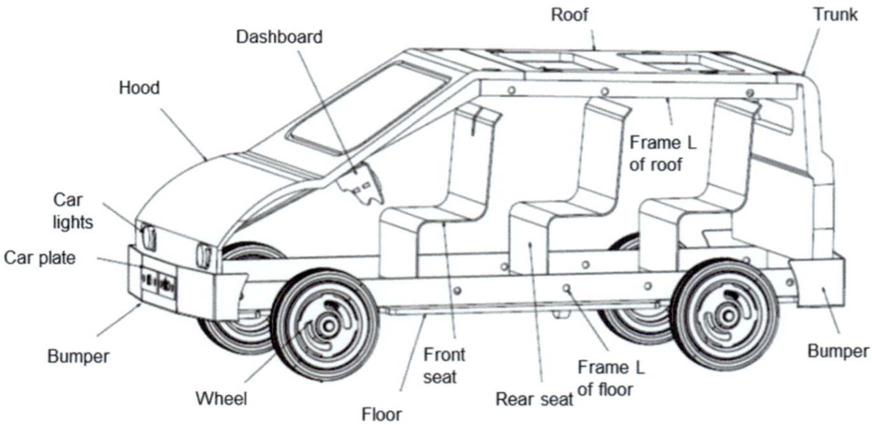

Fig. 5 Lean school car model and its main parts

Initially the students start with a stock of eight finished products, chosen by the production manager among different potential product models. Every two minutes the teacher provides the students with the customer demand (Fig. 8, left column), and the production manager must decide which lot (of four units each) is made, based on timing, customer demand, stock of finished product, WIP, and so on.

After production completion (30 pieces) in a configuration, the students, based on the previous analysis with the head teacher, are responsible for redistributing

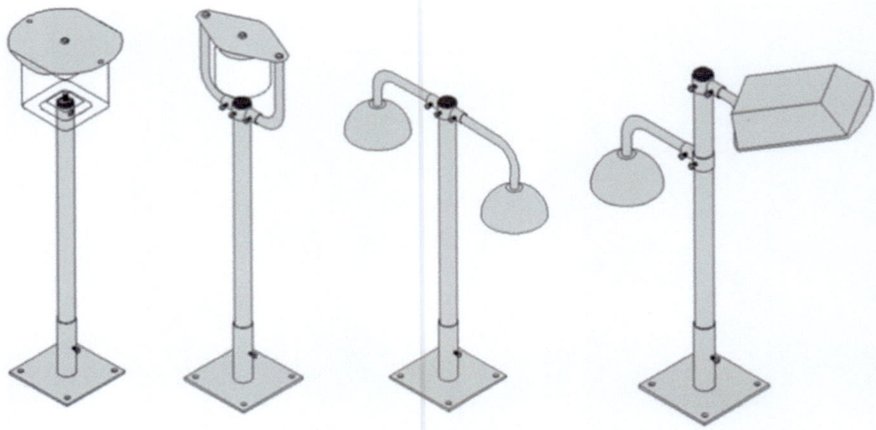

Fig. 6 Various types of lean school streetlights

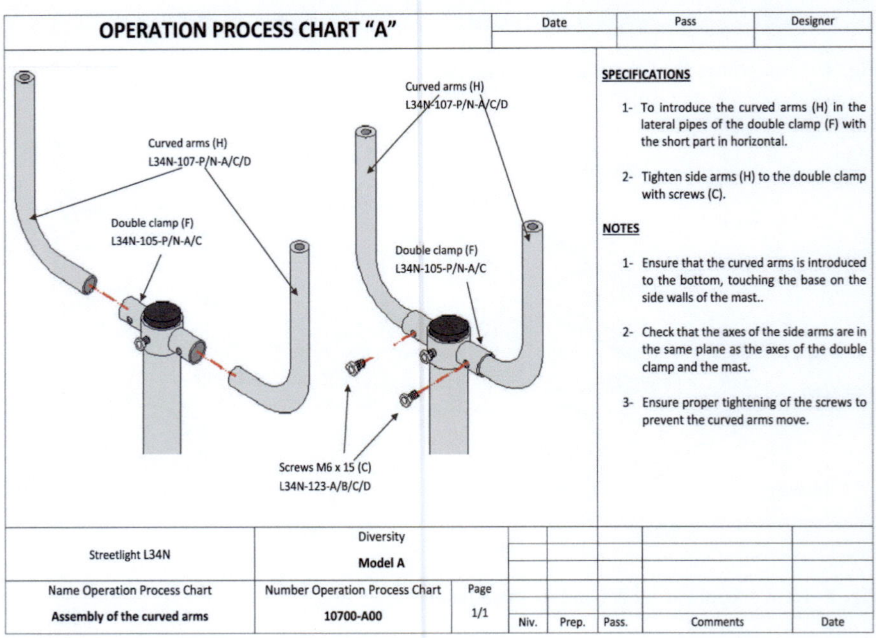

Fig. 7 Operation process chart of a streetlight

workplaces, rebalancing the activities, and proposing corrective actions and design elements that will improve the quality and delivery times.

In the left table of Fig. 8, the professor notes compliance with the customer order. In the right table, he notes the main data used to evaluate the quality of the planning and realization of the production: different times, defects, and costs.

Production 1				
n°	Mod.	Forecast Dealer	Real	Difer.
1	MNC	2	✓	
2	PNO	4	✓	
3	PTC	6	✓	
4	MNO	8	✓	
5	PTO	10	✓	
6	MTC	12	✓	
7	MNO	14	✓+4	
8	PNC	16	✓	
9	MNC	18	✓+4	
10	MNO	20	✓+4	
11	PNC	22	✗	
12	PTO	24	22,7✓	
13	MNC	26	✓	
14	MTO	28	✓+4	
15	MNO	30	✓	
16	MNC	32	✗	
17	PTC	34	33 ✓	
18	MNO	36	✓	
19	MTC	38	✗	
20	MNO	40	✗	
21	PTO	42	34 ✓	+4
22	MNC	44	✗	
23	MNO	46	✗	
24	PNO	48	36,5✓	+4
25	MNC	50	✗	
26	MTC	52	✗	
27	MNO	54	✗	
28	PNC	56	43 ✓	+4
29	MTO	58	48 ✓	+4
30	MNO	60	✓	
Cycle Time				

		PRODUCTION 1		
T	Tack Time (predicted)	2'		
	Cycle Time (real)	3'		
	Vehicles on time	20/30		
	Lead Time			
Q	Internal's defects	A	B	C
		10	1	6
	Customer's defects	A	B	C
		20	6	25
	Reworks (reworked vehicles in line)	7		
	First Time Through (FFT)	2/16		
L	Direct Labour Logistics	3		
	Distance (Steps)	5800		
	WIP+Finished vehicles	8+8		
	Surface m2	300		
C	Direct Labour (Manufacturing + Logistics)	9		
	Coste salarial por vehículo	13,5		
	Internal's defects	A	B	C
	Customer's defects	A	B	C
	Non-quality costs (€/vehicle)	51,6		
	Tools and equipment's cost (per vehicle)	0,015		
	Total Costs = 1+2+3	65,115		

Fig. 8 Planned production and results (*left*), and main data to improve process (*right*)

From the discussion of the results of the first production, the students identify some of the main problems (e.g., quality, response time, costs) and note the need for, among other things, more forklifts, better machinery, more employees, and more stock. Their suggestions are similar to those in real factories. At this time, the professor directs students toward finding the roots of the problems: A3 problem solving, Ishikawa (cause and effect) diagrams, 5 Whys, Pareto charts, histograms, and so on.

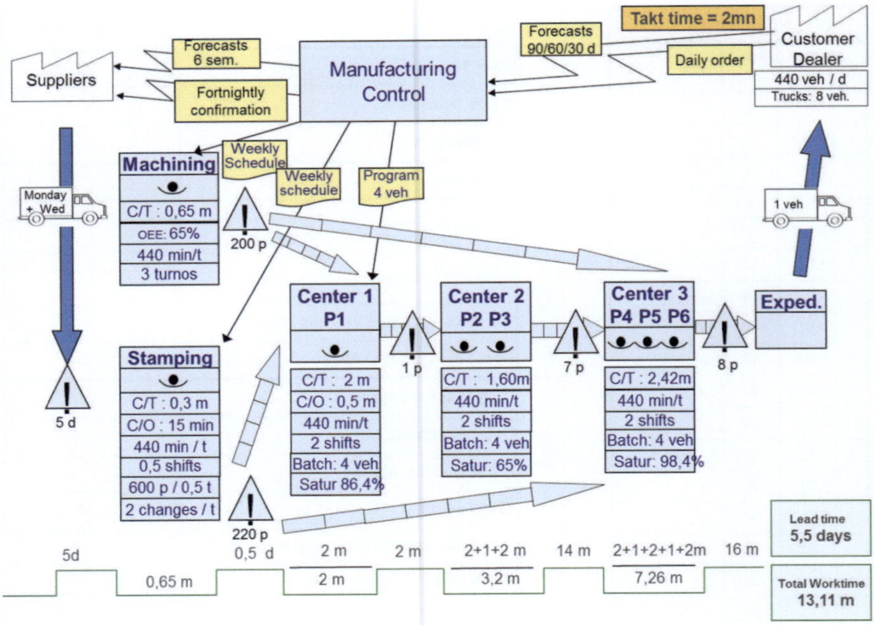

Fig. 9 VSM of the first phase (flow "PUSH" disorganized with low standardization)

To identify the production system elements, the tool Value Stream Management (VSM) is used (Helleno et al. 2015). Based on the results of the Value Stream Analysis (Fig. 9), a catalogue containing means for production optimization can be prepared. In order, to classify the planned measures for optimization, a basic cost-benefit assessment is suitable (Oberhausen and Plapper 2015).

Typical improvement measures are the change of process sequences according to the results of VSM, the prevention of the bottlenecks along the production flow, and quality assurance of products manufactured. Also it is possible increase productivity considering ergonomic aspects such as the location of the tools and work surface height.

Despite being efficient in mapping value-added and nonvalue added activities, VSM is not able to explore and experience the reality of the process. We need to implement all the proposed improvements in the manufacturing process and verify the results.

5 Conclusions

The Lean School is a good example of University-Industry partnership where students learn lean tools in a simulated manufacturing environment, and industrial companies interact with universities to help create world-class engineers and make a significant difference in engineering education.

The Lean School does not focus on teaching manual skills. We use the simulated manufacturing environment to implement the didactic concepts based on learning-by-doing. This laboratory is an innovative training center in Spain, designed for students with

- A unique environment for learning LM, theoretical and practical, very close to the real environment in factories.
- A global comprehension of the management of the operational excellence by the use of different tools: 5S's, Bottleneck Analysis, Continuous Flow, Heijunka, Hoshin Kanri, Jidoka, JIT, Kaizen, Poka-Yoke, SMED, VSM, or Visual Control among others.
- A mentality orientated towards the excellence and the culture of the continuous improvement (Lean Thinking).

Thus, the training (with university students or industrial workers) can be carried out covering the integrity of the aspects of a production system Lean:

- Training on operational tools of Lean (Operating System).
- Training to adapt the organization and management of a LM (managerial structure and management excellence in everyday life).
- Preparation towards a real cultural change (Implication of people).

Also, the Lean School is used today with success both for research purposes as well as for teaching.

References

Bowen, D., & Youngdahl, W. (1998). Lean service: In defense of a production-line approach. *International Journal of Service Industry Management, 9*(3), 207–225.

Bushell, S., & Shelest, B. (2002). Discovering lean thinking at progressive healthcare. *The Journal for Quality and Participation, 25*(2), 20–25.

Dale, E. (1969). *Audio–visual methods in teaching* (3rd ed.), NY: Holt, Rinehart & Winston.

De Zan, G., De Toni, A.-F., Fornasier, A., & Battistella, C. (2015). A methodology for the assessment of experiential learning lean—The lean experience factory case study. *European Journal of Training and Development, 39*(4), 332–354.

Dinkelmann, M., Siegert, J., Bauernhansl, T. (2014). Change management through learning factories in enabling manufacturing competitiveness and economic sustainability. In M. –F. Zaeh (Ed.) (pp. 395–399) Switzerland: Springer International Publishing.

Helleno, A.-L., Simon, A.-T., Papa, M.-C.-O., Ceglio, W.-E., Rossa Neto, A.-S., & Mourad, R.-B.-A. (2013). Integration university–industry: Laboratory model for learning lean manufacturing concepts in the academic and industrial environments. *International Journal of Engineering Education, 29*(6), 1387–1399.

Helleno, A.-L., Pimentel, C.-A., Ferro, R., Santos, P.-F., Oliveira, M.-C., & Simon, A.-T. (2015). Integrating value stream mapping and discrete events simulation as decision making tools in operation management. *The International Journal of Advanced Manufacturing Technology, 80* (5), 1059–1066.

Hines, P., Silvi, R., Bartolini, M., Raschi, A. (2002) A framework for extending lean accounting into a supply chain, in cost management in supply chains. In S. Seuring, M. Goldbach (Eds.) Physica-Verlag.

ILO. (1992). *Introduction to work study* (4th ed.). Geneva: International Labour Office.

International Monetary Fund. (2013). *World economic outlook—Hopes, realities, risks. world economic and financial surveys*. Washington: International Monetary Fund Publication Services.

Jones, D. T., Hines, P., & Rich, N. (1997). Lean logistics. *International Journal of Physical Distribution & Logistics Management, 27*(3/4), 153–173.

Klym, P. (2013). *Lean thinking*. Available via http://www.slideshare.net/klymp000/lean-thinking-26102075

Lamancusa, J.-S., Zayas, J.-L., Soyster, A.-L., Morell, L., & Jorgensen, J. (2008). The learning factory: Industry-partenered active learning. *Journal of Engineering Education, 97*(1), 5–11.

Matt, D.-T., Rauch, E., & Dallasega, P. (2014). Mini-factory—A learning factory concept for students and small and medium sized enterprises. *Procedia CIRP, 17,* 178–183.

Oberhausen, C., & Plapper, P. (2015). Value stream mapping in the "Lean Manufacturing Laboratory". *Procedia CIRP, 32,* 144–149.

Ohno, T. (1988). *Toyota production system*. Productivity Press.

Pozzi, R., Noe, C., & Rossi, T. (2015). Experimenting 'learn by doing' and 'learn by failing'. *European Journal of Engineering Education, 40*(1), 68–80.

Salvador, J., Oliveira, J. (2012). An approach to Foster engineering skills at graduation level: Meeting the industry needs, *iJAC, 5*(2), 20–25.

Seitz, K. F., & Nyhuis, P. (2015). Cyber-physical production systems combined with logistic model—A learning factory concept for an improved production planning and control. *Procedia CIRP, 32,* 92–97.

Veza, I., Gjeldum, N., & Mladineo, M. (2015). Lean learning factory at FESB—University of Split. *Procedia CIRP, 32,* 132–137.

Womack, J. P., Jones, D. T., & Roos, D. (1990). *The machine that changed the world*. New York: Harper Perennial.

Chapter 15
A Review of the Innovative Teaching Activities Carried Out by the School of Industrial Engineering of the University of Valladolid in the Field of Business Organization

Pedro Sanz-Angulo, Juan José de Benito-Martín, Ángel Manuel Gento-Municio, Alfonso Redondo-Castán, José Antonio Pascual-Ruano, Jesús Galindo-Melero and Moisés Javato-Martín

Abstract In the last few years, a group of teachers of the School of Engineering of the University of Valladolid, have been involved in an innovation and continuous improvement process in order to update our teaching activities. On the one hand, we aim to promote the learning process of the existing paradigms in the business organization field. On the other hand, our goal is also that our students can acquire and develop the specific and generic skills that they will need in their future in their professional activity. This endeavour has led us to the implementation of a portfolio of innovation projects e.g. the creation and use of different learning methodologies,

P. Sanz-Angulo (✉) · J.J. de Benito-Martín · Á.M. Gento-Municio · A. Redondo-Castán · J.A. Pascual-Ruano · J. Galindo-Melero · M. Javato-Martín
Department of Business Organization, and Commercialization and Market Research, School of Industrial Engineering, University of Valladolid, Paseo Del Cauce, 59, 47011 Valladolid, Spain
e-mail: psangulo@eii.uva.es

J.J. de Benito-Martín
e-mail: debenito@eii.uva.es

Á.M. Gento-Municio
e-mail: gento@eii.uva.es

A. Redondo-Castán
e-mail: redondo@eii.uva.es

J.A. Pascual-Ruano
e-mail: pascual@eii.uva.es

J. Galindo-Melero
e-mail: jesus.galindo@uva.es

M. Javato-Martín
e-mail: moisesjavato@yahoo.es

C. Hernández (ed.), *Advances in Management Engineering*, Lecture Notes in Management and Industrial Engineering, DOI 10.1007/978-3-319-55889-9_15

the development of tools for supporting teaching activities. This paper is intended to concisely describe the key elements of this innovation effort.

Keywords Teaching innovation · Continuous improvement · Business organization · EHEA · Lean school

1 Introduction

In the course of the last few years several teachers of the Department of Business Organization of the University of Valladolid (UVA), have been establishing an innovation and continuous improvement process in order to promote the learning of certain topics related to industrial organization engineering (e.g. business management, production, logistics, economics, human resources) and to encourage students to develop the skills which will need in the future at a professional and personal level. This endeavour has led us to the development of several tools, methodologies, and documents, which enhance the whole learning course of undergraduate and postgraduate students, and of the students of several degree courses.

All these innovations are driven by the need for a reorientation of teaching activities towards a model which promotes the "knowledge-skills" pairing, in accordance with the change demanded by the job market and encouraged by the European Higher Education Area (EHEA). The objective is that students acquire the knowledge and know-how of their professions (which are related to the content to teach) while they train the key skills (which are related to the way of teaching) for their future performance as professionals (Sanz et al. 2011). In addition to that, these tools have been conceived so that they should be familiar, attractive and easy-to-use, serving a double purpose: on the one hand, motivating students and involving them in the teaching process (because we want them to be active participants, responsible for their own learning); on the other hand, promoting the use of the tools by other teachers or professionals, in the field of organization engineering and in any other knowledge area.

Some of these advances are part of the activities of different teaching innovation projects (TIP), such as the TIP on the development and improvement of the SACC methodology (http://metodologiaaprendizajesacc.blogs.uva.es/) or the recent project "The Lean School: the study and development of different improvement lines for supporting teaching activities and student learning within the EHEA Framework". This paper describes the main elements of these projects, and at the same time presents the initiatives which have been developed by our group in order to improve the teaching and learning processes of its field of knowledge.

2 Teaching Innovation Activities in the Context of the Lean School

In early 2014, Renault Consulting (http://www.renault-consulting.es/) and the University of Valladolid (www.uva.es) inaugurated the first school for Lean learning in Spain (http://escuela-lean.es/), which is located in the School of Industrial Engineering (EII, Escuela de Ingenierías Industriales) of the University of Valladolid. The teaching activities of the school are focused on the removal of any element which does not add value to processes (e.g. production, logistics or service ones), so reducing costs, rising safety and quality standards, improving staff moral and so on; in short, gaining competiveness.

As the former vice-chancellor of the University of Valladolid, Marcos Sacristan, put it: "the school provides the students of industrial engineering and of other degree courses with a piece of training which goes beyond the academic training, and which is mainly a practical, business-oriented one" (EFE Economía 2014). The use of the facilities (see Fig. 1) contributes to give a more practical approach to the subjects which are taught at the EII, as part of several master's degrees (e.g. industrial engineering, logistics) and bachelor's degrees (industrial organization engineering).

In order to achieve the latter, the facilities of the school are initially arranged following the traditional layout of a plant where different car models are manufactured (Fig. 2). During several training sessions, students improve the plant operation by using different Lean Manufacturing (Womack et al. 1992; Womack and Jones 2005) tools and methodologies, such as SMED (Gil et al. 2012), the 5S methodology, Jidoka, Heijunka, standardization, TPM, Poka-yoke, One-piece flow, and so on.

Despite the advantages of the school, we think that there are ways of improving teaching and learning regarding Lean-related concepts. Therefore, we have been developing different upgrading initiatives for teaching and learning; they are based on Lean philosophy and on the implemented training workshops, so profiting from the competitive advantage of having the first Lean School in Spain.

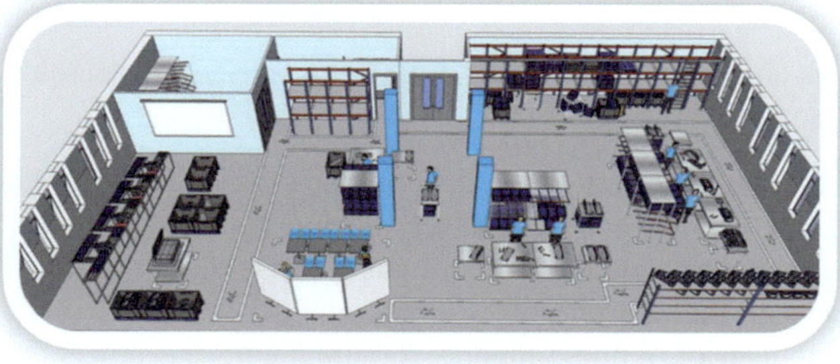

Fig. 1 A 3D model of the Lean School

Fig. 2 Car production at the facilities of the Lean School

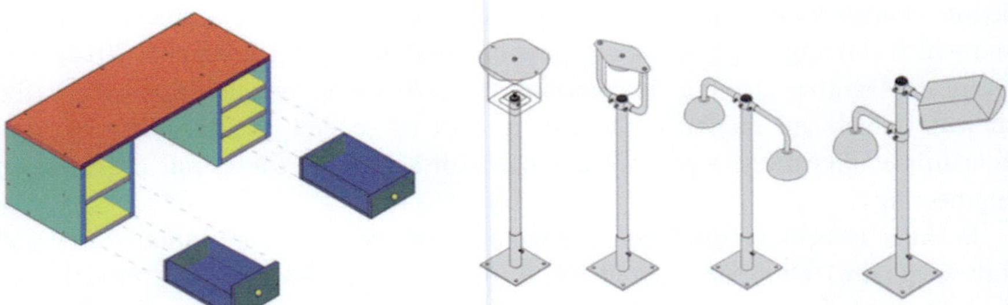

Fig. 3 Examples of new products for the Lean School

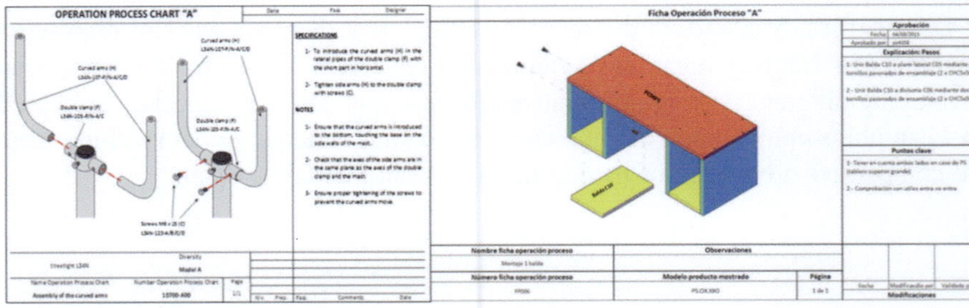

Fig. 4 Examples of some operation-process sheets which have been developed

For instance, we have considered new products-services which could be included in the Lean School, such as: tables, chairs, appliances, lights, and so on. Accordingly, we have defined several designs (Fig. 3) and procedures (Fig. 4), and we are working on the first physical models.

The development of Witness models for simulating the processes of the Lean School (Fig. 5) has been another upgrading initiative. Presently, the modelling of the initial configuration of the Lean School, as well as the one-piece-flow

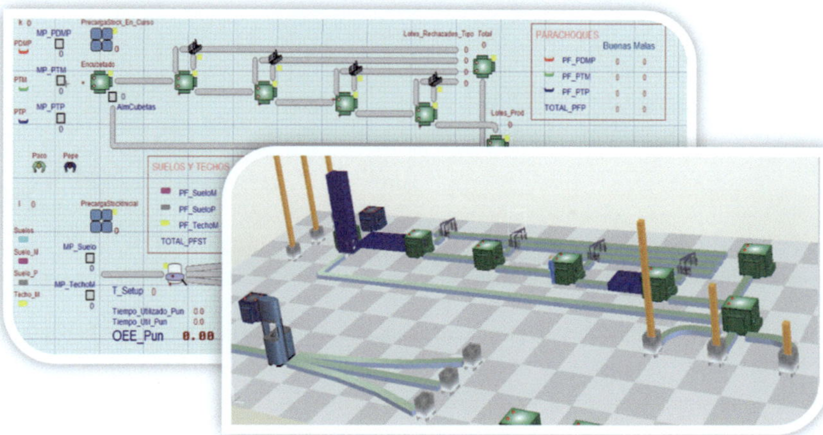

Fig. 5 The simulation model of bumper punching and assembly areas of the Lean School

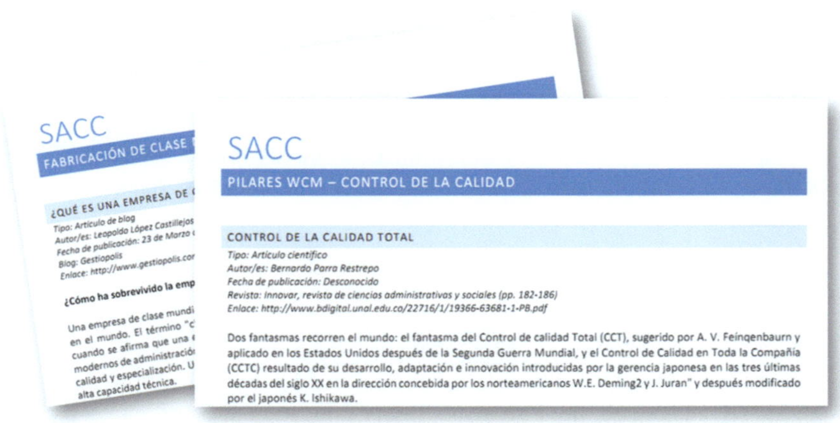

Fig. 6 Supporting documents used in the SACC workshops on Lean tools

configuration, have been completed. These models could be the foundations of future developments, both in the Lean School and elsewhere.

Other improvements involve the use of teaching methodologies and, in particular, the SACC learning methodology (Sanz et al. 2011); in this field, multiple workshops (Fig. 6) have been developed in order to deal with different Lean tools and methodologies.

Finally, we are also working on the use of videogames as a teaching method. In particular, two approaches have been considered: first, the use of Minecraft

Fig. 7 An example of a production plant built using minecraft, in which the 5S methodology has been applied

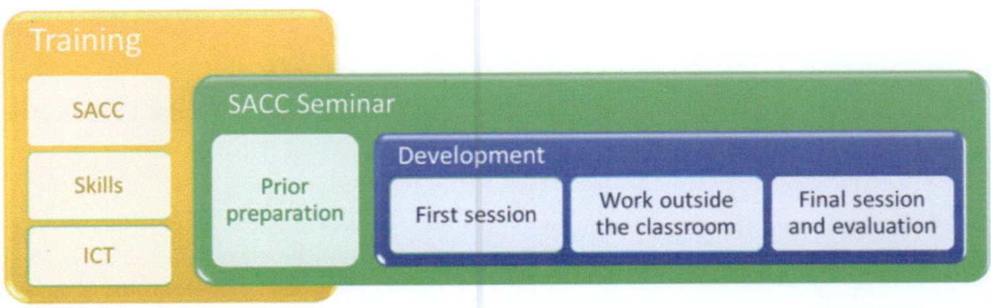

Fig. 8 General vision of the SACC methodology

(https://minecraft.net/) as a way of creating Lean environments (see Fig. 7); secondly, the creation of our own apps and videogames in order to explain some of the Lean tools.

3 SACC Learning Methodology

As previously mentioned, the European Higher Education Area promotes a change in teaching methodologies so that students should acquire the knowledge and know-how of their professions (which are related to the content to teach) while they train the key skills (which are related to the way of teaching) for their future performance as professionals.

The SACC methodology (Workshops on Competitive, Collaborative, Autonomous Learning, *Seminarios de aprendizaje Autónomo, Colaborativo y Competitivo*) was created to achieved this goal by means of combining three styles of learning (autonomous, collaborative and competitive) with the use of Information and Communication Technologies (ICT).

This methodology tries to ensure that students train generic and digital skills, both inside and outside the classroom, while they acquire knowledge in an enjoyable way. In order to do that, the methodology takes into account several key elements, which are depicted in Fig. 8.

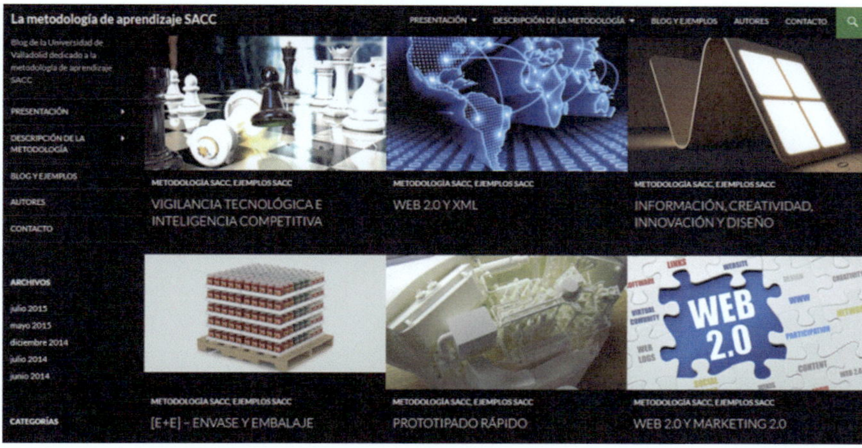

Fig. 9 Website of the SACC methodology

More information on this methodology is available at its website (http://metodologiaaprendizajesacc.blogs.uva.es/). In the latter, its motivation, its key elements, the steps involved and so on, are described, and several application examples to different fields of knowledge are presented (see Fig. 9).

4 Other Activities

During 2013–2014 academic year, the project "A definition of a methodology for the application of Competitive Intelligence and Technological Surveillance to the definition of innovative bachelor's and master's final degree projects within the EHEA framework" was completed. This project involved the development of a Competitive Intelligence Methodology (see Fig. 10) in order to identify the fields of knowledge which are generating innovation and patents (Galindo-Melero et al. 2014). The aim of this methodology is to support teachers when they have to propose bachelor's/master's final degree projects, so that these projects (a) should be focused on to the development of truly innovative, useful solutions (for companies and society) and (b) should allow students to demonstrate their true potential.

Nevertheless, as it was previously mentioned, not all our innovative activities correspond to innovation projects; there are interesting methodologies and software/web tools which have been developed outside this context. That is the case of the LabMacro software, which allows students to improve their analysis and understanding of macroeconomic models (Pascual et al. 2012), so contributing to their more active learning and involving them in their training. This software, which has been developed using Excel and Visual Basic, addresses macroeconomics as a whole: from aggregate demand-supply equilibrium to the complete foreign trade model (Fig. 11).

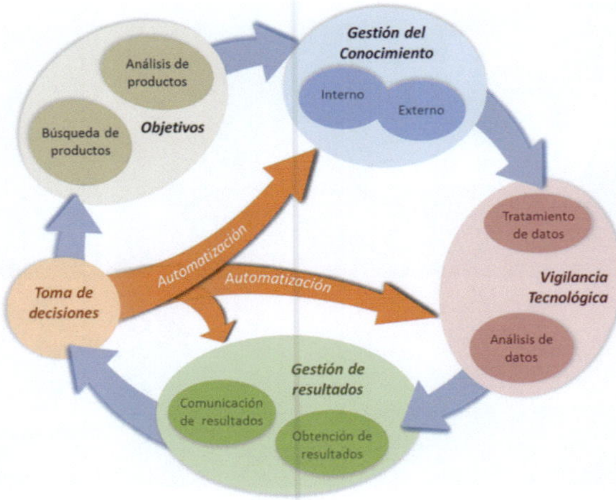

Fig. 10 IC methodology for the definition of bachelor's/master's final degree projects

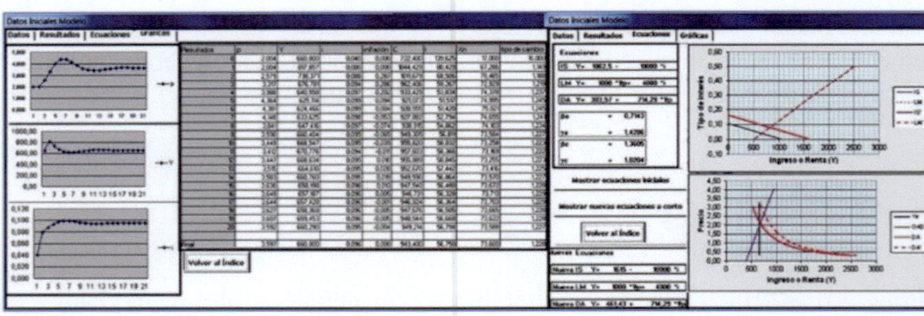

Fig. 11 LabMacro software

As part of the web tools which have been developed, we should highlight a Business Game (www2.eii.uva.es/businessgame) for industrial management training. This business simulator (Fig. 12) allows students to have a "comprehensive view" of a company, including its different functional areas, by means of a method which differs from traditional classes and class notes. The objective of using this software is that students see the consequences of decision making in each area of the company, when competitive conditions are considered, and therefore that they understand the difficulties which may appear in any given time/situation.

On the other hand, a web tool (Fig. 13) for creating practical lectures based on Role Playing has been developed using the experience acquired in the use of this technique in the last 15 years; this tool is hosted at the server of the School of Engineering of the University of Valladolid (www.eii.uva.es/jempresa).

Its goal is that students learn how to negotiate, to lead and manage teams, to make decisions, to organize production and so forth, by means of using the tool, and consequently that they improve their understanding of industrial management

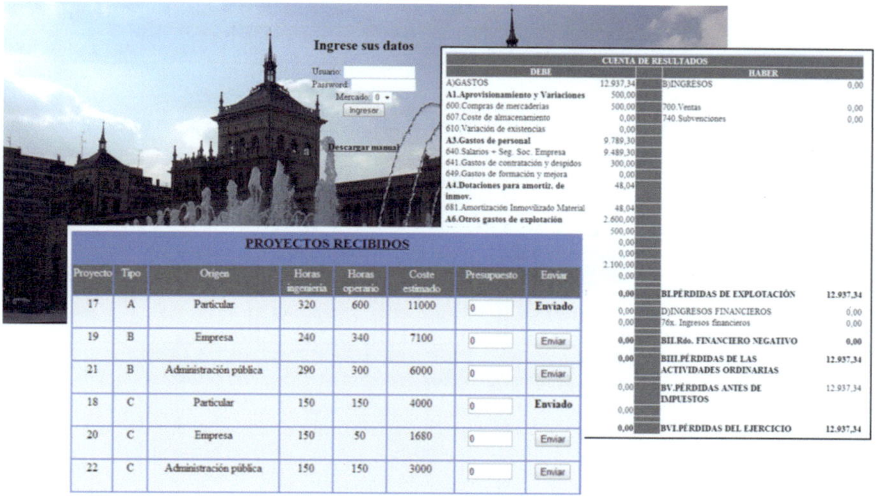

Fig. 12 Web interface which was developed for the business game

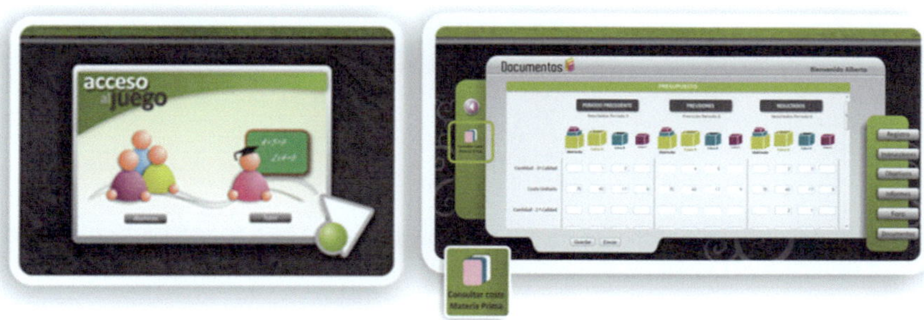

Fig. 13 Role playing website

concepts. Thus, teachers may explain business organizational elements through a role playing method in which each player assumes the role of its character and tries to understand how he thinks and acts.

Lastly, a group dynamics activity for team work training has been developed, profiting from the dynamics which have been implemented in classroom since 2009 (Solana et al. 2012; Redondo et al. 2012). This group dynamics has been conceived in order to strengthen the "Generic Skills" which students should have according the job market. It is the result of collaboration with Mr. Teófilo Solana Miguel (Honorary Associate Professor at our department and former HR training and selection manager at Michelin for 30 years). In this activity, students have to build a crane, using an automatic piece of equipment and by doing so they train skills such as: team work, problem solving, public speaking, effective presentations, and so on.

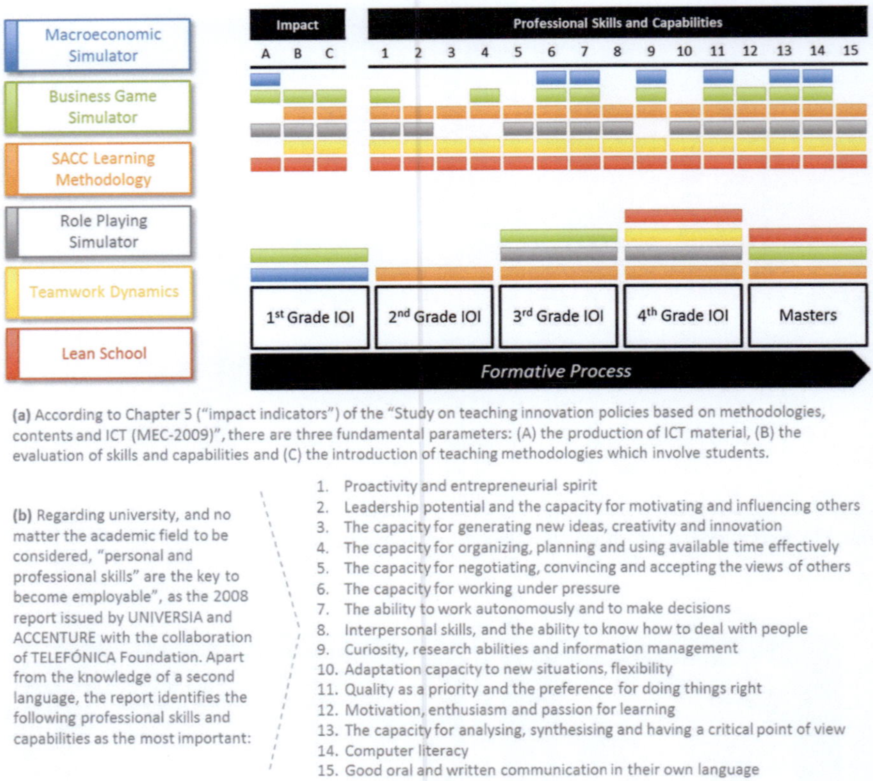

Fig. 14 A summary table of the teaching innovation activities which have been implemented in the field of Industrial Organization

5 Conclusions

A summary table of the improvement activities is depicted in (Fig. 14), which describes the following elements for each action: impact indicators (a), professional skill and capabilities (b) which benefit from each action and their contexts of use within the training process.

Without any doubt, students are the ones who benefit most from these teaching innovation activities. The different tools and methodologies which have been developed contribute to improve student training, both in terms of knowledge and skills (specific and generic ones). Private companies and organizations where the future graduates may work will profit from such activities and so does society. Obviously, the improvement of the learning process also contributes to the satisfaction of teachers, which is also an important outcome, far from a negligible one, as it provides an incentive for day-to-day innovation and improvement.

Last but no least it should be stated that the activities described have not involved a high economic investment all along these years: the development of the tools and methodologies have been mainly achieved by using software which was free, available in an academic version or was purchased by the University of Valladolid; the existing servers of the School of Industrial Engineering, of the University of Valladolid, or those of the department have been used, and so on. However, the activities have required a great deal of human resources investment; without the efforts of the members of the group, not a single achievement would have been possible.

Acknowledgements We would like to begin by expressing our gratitude to all the people who have participated in the development of the different improvement projects and initiatives on which we have been working; to a greater or a lesser extent, they have all made an effort and worked with dedication, contributing not only to the initiatives which have been described in this document but also to others which are not so noticeable. On the other hand, a fact that should be highlighted is the recent recognition of our innovation effort by the Social Council of the University of Valladolid in the latest edition of its Teaching Innovation Awards. We would like to thank them for this recognition, which is has been an incentive for continuing with our teaching innovation efforts and progressing towards a kind of teaching activities which should be adapted to the present needs of society.

References

EFE Economía (2014, January 29). Nace la primera escuela Lean como arma para mejorar la competitividad [News]. Retrieved from http://economia.elpais.com/economia/2014/01/29/agencias/1391001200_277515.html

Galindo-Melero, Sanz-Angulo y De-Benito-Martín. Necesidad de implantar una filosofía metodológica de Inteligencia Competitiva en PYMES y grupos de investigación basada en el análisis de patentes. 9th Iberian Conference on Information Systems and Technologies—CISTI'2014. Barcelona, Spain. 18–21, June 2014.

Gil, M. A., Sanz, P., Benito, J. J., & Galindo, J. (2012). Definición de una metodología para una aplicación práctica del SMED. *Técnica industrial, 298,* 46–54.

Pascual, J. A., Redondo A., & Gento A. M. (2012). Implementación de Prácticas para Incentivar el Estudio de Macroeconomía. En J.C. Prado, J. García, J. A., & Comesaña y A. J. Fernández (Eds.) Industrial Engineering: Overcoming the Crisis. Páginas: 1272 a 1279. Vigo: Reprogalicia, S.L.

Redondo, A., Solana, T., & Pascual, J. A. (2012). Creación y desarrollo de un "Equipo Autónomo de Trabajo": Diseño y construcción de una grúa de papel/cartón. In X. X. Congreso (Ed.), *Universitario de Innovación Educativa en las Enseñanzas Técnicas XX CUIEET.* Las Palmas de Gran Canaria: Escuela de Ingenierías Industriales y Civiles Las Palmas de Gran Canaria.

Sanz, P., de Benito, J. J., & Galindo, J. (2011). La metodología de aprendizaje SACC. Desarrollo práctico de Competencias Transversales y Digitales en los Egresados Universitarios a través de la metodología SACC. *Revista Docencia e Investigación, 21,* 289–308.

Solana, T., Redondo, A., & Pascual, J. A. (2012). Skills training through an independent team. In A. Rodríguez (Ed.) *Innovation and quality in engineering education.* (pp. 81–6). Valladolid: Escuela de Ingenierías Industriales. Universidad de Valladolid.

Universia, Accenture, & Fundación Telefónica. (2008). Las competencias profesionales en los estudiantes preuniversitarios. Madrid: Accenture Ed. Retrieved from http://nstore.accenture.com/acn_com/PDF/Accenture_Las_competencias_profesionales_en_los_estudiantes_preuniversitarios.pdf

Womack, J. P., & Jones, D. T. (2005). Lean thinking: cómo utilizar el pensamiento Lean para eliminar los despilfarros y crear valor en la empresa. Barcelona: Gestión 2000.

Womack, J. P., Jones, D. T., & Roos, D. (1992). *La máquina que cambió el mundo*. Madrid: MacGraw-Hill.

MIX
Papier aus verantwortungsvollen Quellen
Paper from responsible sources
FSC® C105338

If you have any concerns about our products,
you can contact us on
ProductSafety@springernature.com

In case Publisher is established outside the EU,
the EU authorized representative is:
Springer Nature Customer Service Center GmbH
Europaplatz 3, 69115 Heidelberg, Germany

Printed by Libri Plureos GmbH
in Hamburg, Germany